World Sustainability Series

Series Editor

Due to its scope and nature, sustainable development is a matter which is very interdisciplinary, and draws from knowledge and inputs from the social sciences and environmental sciences on the one hand, but also from physical sciences and arts on the other. As such, there is a perceived need to foster integrative approaches, whereby the combination of inputs from various fields may contribute to a better understanding of what sustainability is, and means to people. But despite the need for and the relevance of integrative approaches towards sustainable development, there is a paucity of literature which address matters related to sustainability in an integrated way.

Notes on the quality assurance and peer review of this publication

Prior to publication, the works published in this book are initially assessed and reviewed by an in-house editor. If suitable for publication, manuscripts are sent for further review, which includes a combined effort by the editorial board and appointed subject experts, who provide independent peer-review. The feedback obtained in this way was communicated to authors, and with manuscripts checked upon return before finally accepted. The peer-reviewed nature of the books in the "World Sustainability Series" means that contributions to them have, over many years, been officially accepted for tenure and promotion purposes.

Ayyoob Sharifi · Dahlia Simangan · Shinji Kaneko
Editors

Bridging Peace and Sustainability Amidst Global Transformations

Springer

Editors
Ayyoob Sharifi
The IDEC Institute
Hiroshima University
Higashihiroshima, Japan

School of Architecture and Design
Lebanese American University
Beirut, Lebanon

Shinji Kaneko
The IDEC Institute
Hiroshima University
Higashihiroshima, Japan

Dahlia Simangan
The IDEC Institute
Hiroshima University
Higashihiroshima, Japan

ISSN 2199-7373 ISSN 2199-7381 (electronic)
World Sustainability Series
ISBN 978-981-99-7571-6 ISBN 978-981-99-7572-3 (eBook)
https://doi.org/10.1007/978-981-99-7572-3

This Springer imprint is published by the registered company Springer Nature Singapore Pte Ltd.
The registered company address is: 152 Beach Road, #21-01/04 Gateway East, Singapore 189721, Singapore

Paper in this product is recyclable.

Preface

The interconnections between peace and sustainability are widely acknowledged in academic and policy spheres. The adoption of the 2030 Agenda for Sustainable Development reaffirms this recognition. However, many initiatives addressing peace and sustainability operate independently, overlooking their mutually beneficial relationship. It is essential to integrate peace-building efforts with sustainable development strategies to effectively address complex challenges associated with global transformations at environmental, social, political, and economic levels. This edited volume aims to explore different avenues that bridge the promotion of peace and sustainable development in an integrative manner. It builds upon the previous volume, *Integrated Approaches to Peace and Sustainability*, which was published in early 2023. The first volume successfully drew global attention to the relationship between peace and sustainability.

The opening ceremony of NERPS 2023

The book features a collection of papers that were presented at the NERPS 2023 conference, organized by the Network for Education and Research on Peace and Sustainability. The conference took place from February 28th to March 3rd, 2023 in Thailand's Asian Institute of Technology. Its primary aim was to foster transdisciplinary research and policy discussions on the intersection of peace and sustainability. By providing a platform for dialogue and collaboration among stakeholders focused on these issues, the conference sought to promote further research and initiatives exploring their interconnectedness. With active participation from more than 200 individuals including students, researchers, and practitioners across diverse fields and sectors, a wide range of topics related to peace-sustainability nexus were explored through presentations during this event.

This book delves into the complex relationship between sustainability and peace, providing both theoretical analysis and real-world examples. It recognizes that these concepts are multifaceted and dynamic in nature, constantly evolving to address new challenges. The book explores how sustainability and peace intersect within different social, cultural, economic, and political contexts. Furthermore, it examines the various factors that contribute to the development of sustainable and peaceful societies.

We anticipate that this edited volume will provide valuable insights into the intricate relationship between peace and sustainability. Nevertheless, there is a pressing need for further exploration in order to comprehend the interplay between these two concepts fully. We encourage researchers and policy makers to participate in upcoming conferences and events organized by NERPS where discussions on the peace-sustainability nexus can be expanded upon.

Higashihiroshima, Japan

Ayyoob Sharifi
Dahlia Simangan
Shinji Kaneko

Contents

About the Editors

Ayyoob Sharifi is a Professor at the IDEC Institute, Hiroshima University. He also is a core member of the Network for Education and Research on Peace and Sustainability (NERPS) and has positions in other universities, such as the Lebanese American University. His research is mainly focused on climate change mitigation and adaptation. Ayyoob actively contributes to global change research programs such as the Future Earth and has served as a lead author for global reports such as the Sixth Assessment Report (AR6) of the Intergovernmental Panel on Climate Change (IPCC). The ultimate goal of his education and research activities is to inform actions towards building sustainable and peaceful communities.

Dahlia Simangan is an Associate Professor at the IDEC Institure Hiroshima University. She also holds a position at the Graduate School of Humanities and Social Sciences where she teaches Global Governance and Peacebuilding Case Studies. She is a former Kanagawa University JSPS Postdoctoral Research Fellow (nominated by the United Nations University-Centre for Policy Research in Tokyo). She obtained her Ph.D. in International, Political and Strategic Studies from the Australian National University in 2017, her M.A. in International Relations from the International University of Japan in 2010, and her B.A. in Sociology from the University of the Philippines, Diliman in 2006. Her research interest in peace and conflict studies includes topics on international peacebuilding, the United Nations peacekeeping operations, and International Relations in the Anthropocene. She is the Assistant Editor of Peacebuilding and a member of the Planet Politics Institute.

Shinji Kaneko is the Executive Vice President for Global Initiatives and a Professor at the Graduate School of Humanities and Social Science at Hiroshima University. He graduated from the School of Engineering at Kyushu University majoring in water engineering (Doctor of Engineering). Immediately after the completion of his doctoral program, he joined the Institute for Global Environmental Strategies (IGES) in 1999 as a researcher and conducted research on urban climate policy in Asian megacities for three years. He also worked at Global Change System for Analysis, Research and Training (START) as a Research Fellow for three years from 2005.

In 2018, he was appointed as the Director of Network for Education and Research on Peace and Sustainability (NERPS) at Hiroshima University. He has conducted numerous policy research on natural resources, energy, and the environment in developing countries.

Bridging Peace and Sustainability Amidst Global Transformations

Ayyoob Sharifi, Dahlia Simangan, and Shinji Kaneko

Abstract Researchers, practitioners, and policymakers have long recognized the mutually reinforcing nature of peace and sustainability. This recognition has been reinforced by the unanimous adoption of the 2030 Agenda for Sustainable Development. SDG 16 on Peace, Justice, and Strong Institutions is seen as a catalyst for achieving all other goals. However, despite this growing awareness of the interconnectedness between peace and sustainability, there remains a need to gain a deeper understanding of the specific factors and mechanisms that foster this symbiotic relationship. It is also crucial to examine under what conditions or dynamics we can study such interplay between peace and sustainability. This chapter provides an in-depth analysis of these issues while showcasing how our edited volume contributes to bridging peace and sustainability amidst global transformations.

Keywords Global transformation · Peace · Sustainability · Nexus · Integrated approaches · Global change

1 Introduction

Peace and sustainability have long been recognized by researchers, practitioners, and policymakers as mutually reinforcing. The unanimous adoption of the 2030 Agenda for Sustainable Development was a historical feat, with all United Nations

A. Sharifi (✉) · D. Simangan · S. Kaneko
The IDEC Institute, Hiroshima University, Higashihiroshima, Japan
e-mail: sharifi@hiroshima-u.ac.jp

D. Simangan
e-mail: simangan@hiroshima-u.ac.jp

S. Kaneko
e-mail: kshinji@hiroshima-u.c.jp

A. Sharifi
School of Architecture and Design, Lebanese American University, Beirut, Lebanon

A. Sharifi et al. (eds.), *Bridging Peace and Sustainability Amidst Global Transformations*, World Sustainability Series,
https://doi.org/10.1007/978-981-99-7572-3_1

(UN) member states agreeing on human well-being and ecological balance as urgent and complementary pursuits in this century. SDG 16 on Peace, Justice, and Strong Institutions is considered the enabler of all the other goals. It is a reminder that the outcome of development efforts, such as those in the areas of health, education, gender, and natural resources, will remain short-term and fragile without peaceful conditions and just institutions. But what are the specific factors and mechanisms that facilitate the mutually reinforcing relationship between peace and sustainability? And under what conditions or dynamics can we examine such a relationship? This volume attempts to answer these questions to nuance and pluralize our assumptions behind the peace-sustainability nexus.

We define peace as both the absence of direct conflicts and violence and the presence of conditions to sustain harmony and promote social justice, based on Galtung's typology of negative and positive peace (Galtung 1969). While social justice encompasses the basic needs of survival, well-being, identity and freedom, we also consider ecological balance as one of the fundamental conditions for peace (Galtung 1990). This holistic connotation also applies to our conceptualization of development, which traditionally has been equated with economic growth but has also evolved to mean the richness of human life (UNDP 1994). But it is not just about meeting the needs of the present generation; it is also about ensuring that future generations can meet their needs, or what is now known as sustainable development (UN WCED 1987). In general, sustainability is the ability to maintain existing systems, and sustainable development is an approach toward meeting the present environmental, social, and economic needs of present and future generations. As such, we view sustainability as the capacity of a complex system to adapt to environmental, social, economic, and other types of change or transformation. Nevertheless, our conceptual approach toward understanding peace and sustainability cannot be decoupled from sustainable development. For this reason, this volume takes the Sustainable Development Goals (SDGs) as the starting point for bringing peace and sustainability efforts.

Although a large body of literature and policy documents acknowledge the holistic connotations of peace, studies concerning war and violence are still dominant in peace research, and few studies delve into the mechanisms that promote and sustain peace (Bright and Gledhill 2018). In other words, many of the research agendas on peace do not commence with peace as more than just the absence of conflict (Diehl 2016). This narrow approach could be attributed to the elusiveness of peace as a concept, often regarded as a normative goal rather than an empirical subject (Millar 2021). Other peace-related studies, such as those concerned with climate change, one of the global transformations of our age, also tend to adopt a narrow connotation of peace. These studies often explore the correlation of climate impacts with the emergence or absence of conflicts rather than paying attention to the instances of societal resilience to conflict despite climate-related vulnerabilities (Barnett 2019). In a critical evaluation of the prevailing climate-conflict research agenda, Barnett (2019) advocates for a fundamental shift in perspective, suggesting that peace should not be viewed as an anomaly as it could, in fact, exhibit resilience in the face of climate change. On the other hand, research on sustainability tends to focus on the sustainable extraction of natural resources for human development and environmental protection and conservation, with little attention to intergenerational and adaptability of human-ecosystems coexistence (Salas-Zapata et al. 2018). The reason for this is that sustainability as a concept remains anchored on its predecessor, sustainable development; hence, both terms are often used interchangeably.

Indeed, sustainability, like peace, is an elusive concept, posing challenges to creating and achieving a shared vision of peace and sustainability (White 2013). Rather than proposing or advocating for universal definitions, we argue that holistic, trans-scalar, and nonlinear conceptualizations of peace and sustainability are suitable for examining and dealing with complex global transformations. Doing so minimizes the adverse side-effects of peace promotion on sustainability efforts and vice versa because the drivers, pathways, and outcomes of the peace-sustainability nexus are highly contextualized. For instance, the unregulated environmental footprint of large-scale peacekeeping operations could disrupt livelihood sources, inciting new economic sources of interpersonal violence and societal conflicts. Failure to integrate environmental considerations could further complicate peacebuilding efforts in conflict-affected societies (Conca 2015; UNEP 2022). Similarly, "quick fixes" to climate change impacts, such as massive infrastructural development without sufficient ex-ante assessments, could result in further disenfranchisement of marginalized communities and even environmental degradation, among other types of climate maladaptation (Magnan et al. 2016; Schipper 2022).

The world is confronted by a multitude of issues that undermine peace and sustainability. The ongoing war in Ukraine continues to claim lives, displace civilians, disrupt global food supply, and increase geopolitical tension among major powers. Russia's nuclear threat is not just shifting regional security; it also raises global concern over the potential actions of other nuclear states, such as India, Pakistan, and North Korea. The rise of artificial intelligence (AI) and other digital technologies exacerbates the threats to international peace and security, escalating cyber threats and challenging international norms and regulations. Additionally, global homicides hit a 20-year high, and 108.4 million people remain forcibly displaced worldwide as of the end of 2022 (UN 2023). The long-term impact of the COVID-19 pandemic has also contributed to the challenges in achieving the SDGs, with 48% of the targets moderately or severely off track and 37% having stagnated and regressed (UN 2023). On top of this, a 2022 study shows that human activities transgressed five of the nine planetary boundaries—climate change, biosphere integrity, biochemical flows of phosphorous and nitrogen, land use, and environmental pollutants—making humanity's survival closer to peril (Persson et al. 2022; Steffen et al. 2015). Some of the manifestations of this can be seen in the record-breaking heat and floods that recently hit China, the largest global food supplier. Many countries in the Asia Pacific are still dealing with the aftermath of more severe typhoons/tropical cyclones and floods. Meanwhile, the recent meeting of G20 environmental ministers failed to generate a consensus on GHG emissions despite the continued climate crisis. These compounding challenges have implications across levels—from local communities to global governance—prompting research-driven action to manage the multi-level consequences of global transformations.

This volume contributes to our understanding of these compounding challenges. It builds on the first edition, which consists of research on the interconnections between human societies and environmental sustainability amidst an era of escalating security and ecological crises. The first volume drew on the success of the first NERPS (Network for Education and Research on Peace and Sustainability) Conference, featuring political factors that enhance or prevent the integration of peace and sustainability, case studies of conflict resolution due to the integration of peace and sustainability, and other empirical and normative perspectives for facilitating such integration. This current volume takes stock of these methodologically and ontologically diverse approaches and findings and situates them within ongoing global transformations. The chapters in this second volume demonstrate how an enhanced integration of peace and sustainability components is imperative for addressing complex challenges that come with global transformations that are manifested environmentally, socially, politically, and economically across levels. Collectively, this volume brings to the fore the value of inter-, multi-, and trans-disciplinary research in illuminating the causes and potential solutions to these challenges by bridging the conceptual foundations and practical implications of peace and sustainability.

2 Efforts Toward Building Peace and Sustainability Amidst Global Transformations

The topics covered by the chapters in this volume include exploring the co-benefits and synergies between measures promoting peace and sustainability, examining the role of education in sustainable peacebuilding, analyzing justice frameworks for environmental resource management, and addressing migration challenges within a sustainable development context. Throughout the chapters, it is thoroughly examined how different SDGs intersect with SDG 16 (peace, justice, and strong institutions). These goals include SDG 4 (quality education), SDG 10 (reduced inequalities), SDG 11 (sustainable cities and communities), SDG 12 (responsible consumption and production), SDG13 (climate action), and SDG14 (life below water). The analysis considers the academic perspective while exploring the interconnectedness between these diverse goals with an aim to promote a culture of peace in society.

Chapter 2 analyzed individual-level characteristics and academic researchers' involvement in SDG-related research using Institutional Research data from a university in Japan. The chapter focuses on the relationship between peace and sustainability. The results indicated an overall increase in participation in SDG-related research at the university, accompanied by growth in disciplinary and gender diversity among participants throughout the five-year period (2017–2021). However, despite these improvements, further analysis revealed that disparities based on discipline and gender persisted over time. Considering the findings and limitations of this study, this paper suggests adopting a comprehensive approach that incorporates various types of data (e.g., bibliometric analysis, surveys) to assess researcher diversity more effectively.

Chapter 3 focuses on the relationship between SDG 14 and SDG 16. The authors conducted a review of the literature and engaged researchers and policy makers through an interactive workshop to gain insights into this connection. According to the literature, a key benefit is that more inclusive decision-making processes can lead to improved ocean and coastal governance by promoting knowledge sharing and compliance. However, there is limited research available on the synergistic interactions between SDG 14 and 16. Nonetheless, existing evidence suggests that fostering peace can contribute to the preservation of ocean ecosystems, while effective ocean governance helps mitigate risks of violent conflicts. On the other hand, stricter regulations pertaining to marine resources might increase competition for these resources, potentially leading to conflicts.

Interactions between SDG 4 and other SDGs (e.g., 13 and 16) are examined in Chap. 4. The chapter endeavors to advance the integration of environmental education within Islamic religious education using a creative imagination approach. Given Indonesia's current profound ecological crisis, educators are under scrutiny to develop more meaningful and impactful education, particularly in the realm of Islamic religious instruction. The findings demonstrate that environmental education is effectively incorporated into pertinent topics, specifically within the context of Islamic education through utilizing creative imagination strategies. The investigation reveals that an educational model promoting love for the environment can be applied by employing ecological education principles infused within Islamic learning frameworks bolstered by creative imagination techniques.

Chapter 5 delves into the significance of dignity and its connection to peacebuilding. The authors aim to shed light on caste-based discrimination as a fundamental obstacle to fostering social cohesion, which in turn undermines efforts towards peacebuilding at a societal level in Nepal. Particularly focused on understanding how adolescent girls perceive dignity, the findings unequivocally confirm that caste-based discrimination violates their sense of dignity. Consequently, these young girls actively avoid interactions with individuals from higher castes and express anger towards them, thus impeding social cohesion within society. Given that caste-based inequality serves as a source of conflict, it is imperative that such issues be prioritized within development agendas aiming to promote social coherence effectively. Notably, any violation or neglect of human dignity has far-reaching consequences: it not only sparks conflicts but also disrupts peace and obstructs sustainable development initiatives. Therefore, achieving desired outcomes in terms of societal-level peaceful coexistence necessitates considering dignity as an essential element within all endeavors pertaining to peacebuilding.

Chapter 6 critically analyzes the role of youth leadership in peacebuilding programs. This chapter specifically focuses on the development assistance initiatives in the Philippines that have successfully promoted youth participation within the Bangsamoro Autonomous Region for Muslim Mindanao (BARMM). Recognizing and highlighting how crucial youth engagement is to fostering sustainable peace, the chapter aims to thoroughly examine and evaluate the active involvement and significant contributions made by various Bangsamoro youth civil society organizations towards achieving lasting peace in Mindanao. Through an institutional approach, the authors strive to identify key factors that create a conducive environment for young individuals to become proactive agents of positive change through effective peacebuilding efforts. The findings from this investigation strongly suggest that establishing formal platforms for meaningful participation alongside access to diverse funding sources significantly enhances and sustains robust engagement between youth organizations operating within BARMM structures. Furthermore, forging strong networks among different Bangsamoro youth groups has been found instrumental towards promoting broad-based participation across all levels; however, it is important to address intergenerational disparities that can pose challenges for ensuring full-fledged involvement of young people in these endeavors.

Chapter 7 explores the significant role of the Melanesian Spearhead Group (MSG), a sub-regional organization from the Pacific Islands, in conflict management. The author asserts that despite its relatively small size, the MSG functions as a credible and effective actor in addressing the West Papuan conflict. This chapter analyzes how the MSG utilizes preventive diplomacy strategies to manage this conflict and highlights their various contributions towards promoting positive engagement and minimizing escalations between West Papua and Indonesia. Despite recurring challenges, these efforts have proven crucial in maintaining peace between both parties.

In Chap. 8, the topic of justice in strategic environmental assessment is discussed. The Vientiane Transport Master Plan aims to create a sustainable urban transport system in Laos by 2040 with inclusiveness as its vision. As part of this plan, a Participatory Strategic Environmental Assessment Scheme has been developed to ensure the sustainability of the urban and transport system. The authors argue that the participatory approach has proved effective in identifying and addressing transboundary issues.

Chapter 9 explores the examination and articulation of different dimensions of environmental justice in urban green spaces in the Global South. The study reveals that most articles primarily focus on distributional justice, while less attention is given to procedural and recognition dimensions. Based on these findings, a conceptual framework has been developed to provide guidance on key criteria for creating inclusive and sustainable green spaces. This framework connects the concept of environmental justice with SDG 11 and SDG 16, highlighting the significance of promoting public participation and advocating for transparent decision-making processes that include diverse perspectives.

Chapter 10 focuses into resource flow sustainability in Ghana. The authors highlight a significant correlation between Domestic Material Consumption (DMC), GDP per capita, and material intensity (DMC/capita) over the course of four decades. They express concerns regarding inadequate resource management and its potential consequences for environmental sustainability.

Chapter 11 delves into the crucial issue of promoting and safeguarding the welfare of migrant workers, highlighting the concerning oversight by the Taiwanese government in recognizing their significance. Migrant workers endure substantial financial burdens due to the prevailing brokerage system, an aspect that cannot be ignored.

Although Taiwan undoubtedly benefits from migrant workers' contributions, it is evident that more must be done to ensure their protection. Accordingly, it is imperative for Taiwan to align with a global vision that prioritizes a recruitment system benefiting both local economic needs while mitigating any potential risks of exploitation faced by migrant workers.

Chapter 12 also focuses on migrants based on insights obtained from highly skilled Thai migrant workers in Japan. The analysis reveals that despite pressure to conform to Japanese norms and values, Thai participants maintain a strong sense of self-continuity. The study has implications for future identity research that considers cultural and anthropological perspectives.

3 The Way Forward for Bridging Peace and Sustainability

The first volume of this series mainly focused on issues related to politics and governance, refugees, media, women, culture, and religion as shown in Fig. 1a (Sharifi et al. 2023). This volume expands the coverage of the peace-sustainability nexus, as depicted in Fig. 1b world cloud of terms related to conflict resolution, peacebuilding, justice, youth and female empowerment, social dignity, and resource management, among other topics. The contributions in this volume are just a subset of a myriad of peace and sustainability issues, and more issues could be magnified in other contexts and scales. As global transformations—from climate change to digital revolution—continue to generate interconnected issues undermining peace and sustainability, we call for enhanced transdisciplinary research collaborations that can inform various activities of different actors across levels of governance. While the research presented in this volume contributes to a better appreciation of the connection between peace and sustainability, there is still much work to be done in order to fully grasp its dynamics. Empirical and conceptual research into other topics, such as health, ethics, and economy, as emphasized by Sharifi et al. (2021), will further enrich our understanding of this complex relationship. It is crucial to extend this examination toward multiple pathways between the SDGs in order to better situate the conditions and dynamics in which peace and sustainability can be mutually reinforcing. The slow progress in meeting the SDGs, as evaluated by the UN (2023), prompts a continued examination of these pathways.

Fig. 1 The overall thematic focus on this volume (**a**) and the previous volume titled "Integrated Approaches to Peace and Sustainability" (**b**)

References

Barnett J (2019) Global environmental change I: climate resilient peace? Prog Hum Geogr 43(5):927–936. https://doi.org/10.1177/0309132518798077

Bright J, Gledhill J (2018) A divided discipline? Mapping peace and conflict studies. Int Stud Perspect 19(2):128–147. https://doi.org/10.1093/isp/ekx009

Conca K (2015) An unfinished foundation: the United Nations and global environmental governance. Oxford University Press

Diehl PF (2016) Exploring peace: looking beyond war and negative peace. Int Stud Quart 60(1):1–10. https://doi.org/10.1093/isq/sqw005

Galtung J (1969) Violence, peace, and peace research. J Peace Res 6(3):167–191. https://doi.org/10.1177/002234336900600301
Galtung J (1990) Cultural violence. J Peace Res 27(3):291–305
Magnan AK, Schipper Elf, Burkett M, Bharwani S, Burton I, Eriksen S, Gemenne F, Schaar J, Ziervogel G (2016). Addressing the risk of maladaptation to climate change. WIREs Climate Change 7(5):646–665. https://doi.org/10.1002/wcc.409
Millar G (2021) Ambition and ambivalence: reconsidering positive peace as a trans-scalar peace system. J Peace Res 58(4):640–654. https://doi.org/10.1177/0022343320941909
Persson L, Carney Almroth BM, Collins CD, Cornell S, de Wit CA, Diamond ML, Fantke P, Hassellöv M, MacLeod M, Ryberg MW, Søgaard Jørgensen P, Villarrubia-Gómez P, Wang Z, Hauschild MZ (2022) Outside the safe operating space of the planetary boundary for novel entities. Environ Sci Technol 56(3):1510–1521. https://doi.org/10.1021/acs.est.1c04158
Salas-Zapata WA, Ríos-Osorio LA, Cardona-Arias JA (2018) Knowledge, attitudes and practices of sustainability: systematic review 1990–2016. J Teach Educ Sustain 20(1):46–63. https://doi.org/10.2478/jtes-2018-0003
Schipper ELF (2022) Catching maladaptation before it happens. Nat Climate Change 12(7), Article 7. https://doi.org/10.1038/s41558-022-01409-2
Sharifi A, Simangan D, Kaneko S (2021) The literature landscape on peace–sustainability nexus: a scientometric analysis. Ambio 50(3):661–678. https://doi.org/10.1007/s13280-020-01388-8
Sharifi A, Simangan D, Kaneko S (2023) Integrated approaches to peace and sustainability. Springer Nature Singapore. https://doi.org/10.1007/978-981-19-7295-9_1
Steffen W, Richardson K, Rockström J, Cornell SE, Fetzer I, Bennett EM, Biggs R, Carpenter SR, de Vries W, de Wit CA, Folke C, Gerten D, Heinke J, Mace GM, Persson LM, Ramanathan V, Reyers B, Sörlin S (2015) Planetary boundaries: guiding human development on a changing planet. Science 347(6223):1259855. https://doi.org/10.1126/science.1259855
UN WCED (United Nations World Commission on Environment and Development) (1987) Our common future—Brundtland report (UN Doc. A/42/427). United Nations World Commission on Environment and Development. https://sustainabledevelopment.un.org/content/documents/5987our-common-future.pdf
UNDP (United Nations Development Programme) (1994) Human development report 1994: new dimensions of human security. Oxford University Press. http://www.hdr.undp.org/en/content/human-development-report-1994
UN (United Nations) (2023) The sustainable development goals report 2023: special edition. https://unstats.un.org/sdgs/report/2023/
UNEP (United Nations Environment Programme) (2022) Greening the Blue Report 2021: the UN system's environmental footprint and efforts to reduce it. United Nations. https://doi.org/10.18356/9789210011129
White MA (2013) Sustainability: I know it when I see it. Ecol Econ 86:213–217. https://doi.org/10.1016/j.ecolecon.2012.12.020

Toward Synergies Between Peace and Sustainability: Using Institutional Research Data to Explore the Diversity of Participants in SDG-Related Research

Megumi Watanabe

Abstract Complex interconnections exist between the Sustainable Development Goals (SDGs), which is exemplified by the Peace and Sustainability Nexus. While this interconnectedness makes SDGs initiatives challenging, it also indicates synergies among the goals. Realizing potential synergies requires promoting collaboration among researchers with different perspectives and expertise. However, there is a lack of geographic and institutional diversity among contributors to SDG-related research, and little is known about the diversity of individual researcher characteristics such as discipline and gender. Most previous studies on trends in SDG-related research analyzed publication data from academic databases (e.g., Web of Science and Scopus) with limited author profile information. Using Institutional Research (IR) data from a university in Japan (including both faculty publication data and demographic data), this study examined individual-level characteristics and a variety of academic researchers who participated in SDG-related research. The results showed that the disciplinary and gender diversity of research participants generally increased over the five-year period (2017–2021), along with an increase in overall participation in SDG-related research at the university. Despite the increase in diversity, bivariate and multivariate analyses of participation likelihood revealed that differences by discipline and gender remained significant over time. Based on the findings and limitations of this study, this paper discusses the need for a more comprehensive approach to examining researcher diversity using a variety of data (e.g., bibliometric, IR, and survey data).

Keywords SDGs · academic researchers · Diversity · Discipline · Gender

M. Watanabe (✉)
Office of University Strategy, Hiroshima University, 1-3-2 Kagamiyama Higashihiroshima, Hiroshima 739-8511, Japan
e-mail: megumiw@hiroshima-u.ac.jp

A. Sharifi et al. (eds.), *Bridging Peace and Sustainability Amidst Global Transformations*, World Sustainability Series,
https://doi.org/10.1007/978-981-99-7572-3_2

1 Introduction

Complex barriers hinder the sustainable development of humans, society, and the world. The complexity of sustainability challenges is demonstrated by the interconnectedness between the United Nations Sustainable Development Goals (SDGs). One example of this interconnectedness is the interaction between climate action (SDG13) and peace (related to SDG16). Numerous studies have shown that disasters caused by climate change are linked to conflict and violence, undermining peace (see Sharifi et al. 2021 for a review of the literature and a bibliometric analysis). This interconnectedness between climate change and peace suggests that peace promotion must be integrated with initiatives in environmental sustainability (Nicoson 2021; United Nations 2012). At Hiroshima University (HU), a group of researchers pioneered a new field called the "Peace-Sustainability Nexus" to facilitate transdisciplinary research and policy dialogue on the intrinsic interlinkages between peace and sustainability (Network for Education and Research on Peace and Sustainability [NERPS] n.d.; Sharifi et al. 2023; Simangan et al. 2021).

As illustrated by the development of the peace-sustainability nexus, the interconnectedness between SDGs may lead to synergies among goals. To achieve potential synergies, it is recommended that a holistic approach to the SDGs be taken by inclusive groups of researchers representing different nationalities, genders, and disciplines (Elsevier 2020). However, SDG-related research lacks diversity in terms of geographic location (e.g., regions and countries) and researchers' institutional affiliations (Elsevier 2020; Meschede 2020; Salvia et al. 2019; Sharifi et al. 2021; Sweileh 2020). While a gender imbalance among influential researchers has been noted (Sharifi et al. 2021), little is known about gender diversity among participants in SDG-related research (Khalikova et al., 2021). Regarding the diversity of disciplines, studies have examined the research areas of SDG-related articles (Meschede 2020) and the subject areas of the journals in which articles were published (Sweileh 2020). However, these studies did not examine the researchers' areas of expertise at the individual level.

Considering the argument for more inter- and transdisciplinary collaboration in SDG-related research (Elsevier 2020; NERPPS n.d.), the gap in the literature regarding researchers' disciplinary diversity is both surprising and problematic. In addition, considering the growing importance of incorporating a gender perspective into SDGs (Herbert et al. 2022), it is essential to analyze the diversity of researchers through a gender lens. Therefore, this study aims to explore the diversity of academic researchers participating in SDG-related research, focusing on various individual-level characteristics, particularly discipline and gender. I used Institutional Research (IR) data from Hiroshima University, which matched the basic demographic data of faculty members with publication data extracted from academic databases.

2 Literature Review

Why should we pay attention to the diversity of researchers engaged in the SDGs? This is because the diverse perspectives and expertise collected within a group can help solve complex challenges and barriers (Hill et al. 2014; Page 2007, 2017). Researchers from universities play a central role in efforts toward achieving the SDGs on a global scale, not only through the production and dissemination of new knowledge but also through human resource education and engagement with various stakeholders (Purcell et al. 2019). Addressing diversity among academic researchers can be a pathway to potential synergies among the SDGs and the nexus between peace and sustainability (Elsevier 2020; Kasuga 2021; Sharifi et al. 2020).

2.1 The Role of Diversity in Groups

Previous research in non-academic settings suggests that heterogeneous groups are more likely to find innovative ideas and solutions than homogeneous groups. This occurs despite the difficulties that increased diversity may bring (e.g., communication problems, conflict, and difficulty in implementing ideas)[1] (e.g., DiTomaso et al. 2007; Herring 2009; Hundschell et al. 2021; Mannix and Neale 2005; McLeod et al. 1996; van Knippenberg and Schippers 2007). In academic cases, there are mixed findings depending on the type of diversity and the context (De Saá-Pérez et al. 2015). Gender diversity is associated with scientific innovation and more impactful ideas (Yang et al. 2022). In addition, Freeman and Huang (2015) found that ethnically and geographically diverse research teams make greater scientific contributions than less diverse teams; their study measured the scientific contributions by publications in high-impact journals and the number of citations. Moreover, especially in Earth Science, papers written by international teams are cited more often than those written by single-country teams, whereas papers written by racially/ethnically heterogeneous teams receive fewer citations than those written by racially/ethnically homogeneous teams (Lerback et al. 2020).

Overall, existing research shows the positive effects of diversity on group outcomes in both academic and non-academic settings. It is important to keep in mind that information diversity and demographic diversity (i.e., diversity in social characteristics such as gender, race, and nationality) are different concepts (Mannix and Neale 2005; Phillips et al. 2011). Informational diversity is defined as "the differences in information, opinions, perspectives and modes of thought and action that are relevant for the task at hand being completed by a group" (Phillips et al. 2011, p. 255),

[1] To overcome difficulties associated with increased diversity in teams, fostering inclusive cultures and practices is important (Hill et al. 2014; Lerback et al. 2020). In academia, for example, good team management, including support for researchers to understand the values and practices of other disciplines, increases the effectiveness of interdisciplinary collaboration (Bozeman et al. 2016).

and is influenced by various factors, including educational and occupational experience. This study uses demographic diversity as a proxy for information diversity because social identity influences people's lived experiences (Page 2017).

Given the empirically supported benefits of diversity in groups, it will be useful to better understand the current state of diversity among researchers participating in SDG-related research and take measures to increase diversity accordingly. In the next section, we review what we already know (and do not know) about this issue.

2.2 Diversity of Researchers Contributing to the SDGs

Several studies have examined research trends related to SDGs (e.g., growth in publications, thematic patterns, and international collaboration) (e.g., Meschede 2020; Sharifi et al. 2021). Although diversity was not the main theme in most of these studies, the results indicate imbalances in research contributions at the institutional, national, and regional levels. At the institutional level, certain prominent institutions, such as the World Health Organization, London School of Hygiene and Tropical Medicine, and Harvard University, are more active in producing SDG-related articles than others (Sweileh 2020). At the country level, SDG-related articles are overwhelmingly from high-income countries such as the United States and the United Kingdom, although the degree varies by SDG (Elsevier 2020; Meschede 2020; Sharifi et al. 2021; Sweileh 2020). At the regional level, European countries produce the largest share of SDG-related articles, whereas Eastern Mediterranean countries produce the least (Sweileh 2020). Furthermore, the well-studied SDGs differ by region and continent (Meschede 2020; Salvia et al. 2019; Sweileh 2020).

In contrast, the diversity of SDGs contributors has not been adequately studied at the individual researcher level. This may be due to data limitations in previous studies. While some studies collected data through surveys (Khalikova et al. 2021; Salvia et al. 2019), most conducted bibliometric analyses using publication data obtained from academic databases such as the Web of Science and Scopus (e.g., Meschede 2020; Sharifi et al. 2021; Sweileh 2020). The limited information on author profiles in academic databases makes it difficult to analyze author characteristics. For example, when gender information is unavailable, it is assigned by inference from names (e.g., Larivière et al. 2013; Liu et al. 2020).

Inspired by the existing knowledge and the lack thereof, this study explored the diversity of researchers participating in SDG-related research at an individual level. This article presents the case of HU, where the author works. By utilizing IR data on the university faculty, I report on the demographic characteristics and variety of researchers who participated in SDG-related research by publishing journal articles. In addition to the availability of IR data, there are reasons why HU is a good case study and why sharing HU's experiences will contribute to the literature. As a research university founded in Hiroshima, the first city in the world to be exposed to the atomic bomb, HU is strongly committed to SDGs, especially the pursuit of peace and sustainability. In October 2017, HU announced its long-term vision, SPLENDOR PLAN

2017 (SPLENDOR stands for "Sustainable Peace Leader Enhancement by Nurturing Development of Research"). The SPLENDOR PLAN embodies the university's guiding principles as an action plan, aiming to establish "Science for Sustainable Development." The action plan was in line with the SDGs proposed by the United Nations in 2015. Through the release of the SPLENDOR PLAN, HU reaffirmed its commitment to contributing to the well-being of humans, society, and the future. It would be valuable to identify possible imbalances in faculty participation in SDG-related research at universities publicly committed to promoting SDGs. The findings of this study have implications not only for HU, but also for other universities striving to promote SDGs.

3 Methods

As part of efforts to promote the SDGs, HU has created scales to measure the extent to which its faculty contributes to each of the first 16 SDGs through publications. These scales of publications were created by matching two types of datasets: the HU and SDG article datasets.

HU regularly extracts data on articles (in all languages) authored by affiliated faculty from the Web of Science and Scopus. The extracted article list is matched to each HU author's basic demographic data kept by the university. HU began compiling these matched lists of articles and author information in 2012 for IR and strategic planning (hereafter, these lists are called *the HU Article Dataset*). Once a year, the IR Office requests that the faculty check information on the articles they published in the previous year through the university's online portal. This confirmation process improves the accuracy of the HU dataset. Meanwhile, from the list of publications mapped to each SDG in Scopus (Rivest et al. 2021), those that included HU faculty among the authors were extracted (hereafter referred to as *the SDG Article Dataset*). Then, the HU and SDG Article Datasets were matched to combine information on (a) which HU faculty members wrote the article, (b) to which SDG(s) each article was mapped, and (c) the HU authors' basic demographic characteristics.

Based on information from the HU and SDG Article Datasets, the IR Office calculates the scales of SDG publications that indicate the extent to which each HU faculty member contributes to each of the first 16 SDGs each year. A score of one was given to each article. Currently, scales have been created for publication years 2012–2021. This study utilized these scales to analyze the diversity among HU faculty members who participated in SDG-related research. The author received de-identified data after the IR Office performed data processing. This study was approved by the Ethics Review Committee of the Research Institution for Higher Education at HU.

4 Results and Discussion

4.1 Overall Growth

Figure 1 presents the number of HU faculty members who published articles related to *at least one of the first 16 SDGs* in the publication years 2012–2021, and Fig. 2 presents the number of HU faculty members who published articles related to *two and more of the first 16 SDGs* during the same period. Both figures show a growth in the number of faculty members who published articles on SDG-related research over time, with a noticeable increase after 2017, when HU announced the SPLENDOR PLAN 2017. Between 2017 and 2021, the number of HU faculty members who published articles related to at least one SDG increased by a factor of 1.5, from 361 to 558 (see the blue and red bars in Fig. 1). Among the faculty that published articles related to at least one SDG, an increasing percentage published articles about two or more SDGs (see the added orange line in Fig. 2). The HU faculty's increasing tendency to conduct and publish research related to multiple SDGs may reflect their interest in bridging the different foci of SDGs. However, the data used here did not include faculty attitudes toward research, and it was not possible to investigate who was willing and able to contribute to the realization of nexuses between SDGs. From this point on, the results of the analyses are discussed by comparing two time points, 2017 and 2021, at the beginning and end of the five-year period, in which HU faculty members were increasingly active in SDG-related research.

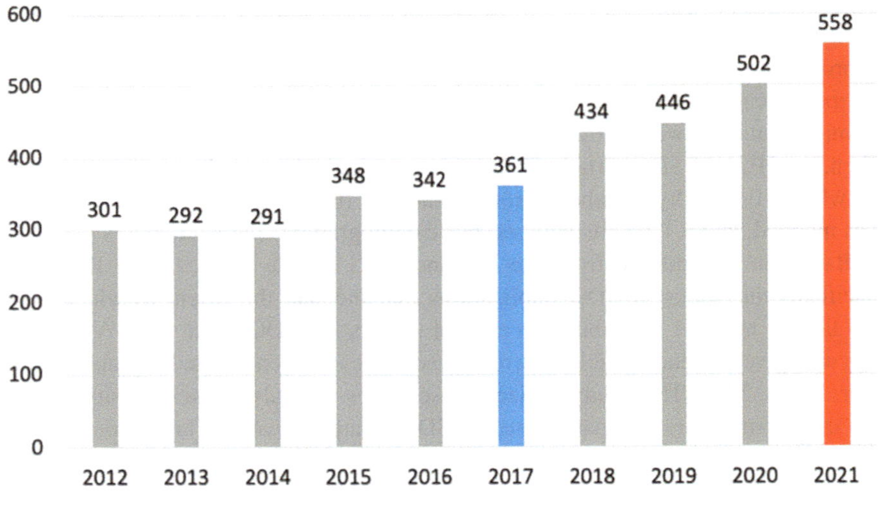

Fig. 1 Transition of the number of HU faculty who published articles related to at least one SDG (2012–2021)

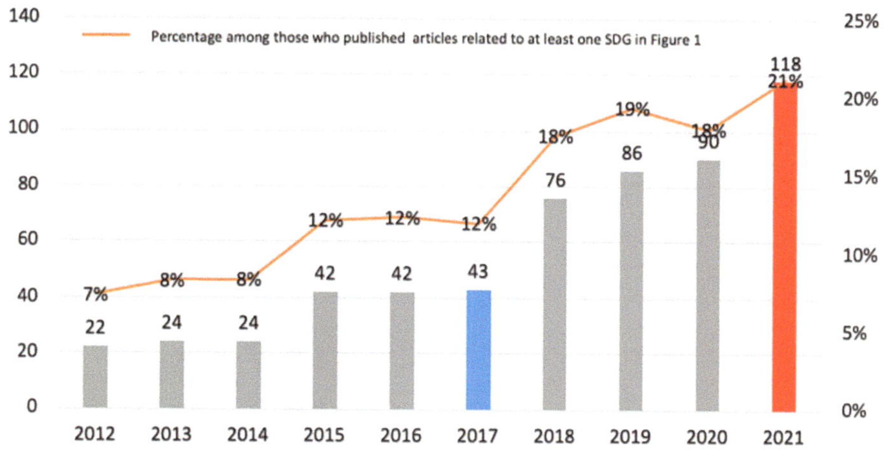

Fig. 2 Transition of the number of HU faculty who published articles related to two and more SDGs (2012–2021)

4.2 *Breakdown by SDG*

Figure 3 shows that, between 2017 and 2021, the number of HU faculty members who published articles related to the SDGs increased for all 16 SDGs. SDG3, Good Health and Well-Being, was the most researched SDG in 2017 and 2021. SDG3 tends to be studied globally, especially in Africa, the Eastern Mediterranean, and Southeast Asia (Elsevier 2020; Meschede 2020; Sweileh 2020).

Figure 4a, b illustrate the discipline representation of the HU faculty whose published articles were mapped to each SDG in 2017 and 2021 (see Appendix A for the discipline categorization). More disciplines are represented in the SDGs in 2021 than in 2017. For example, in 2021, the Arts and Humanities category was represented in seven SDGs, more than doubling the three SDGs in 2017. Health Sciences is represented in 12 SDGs in 2021 and a significant increase from three SDGs in 2017.

Figure 5a, b indicate the gender representation of the HU faculty whose published articles were mapped to each SDG in 2017 and 2021. Women published articles related to all SDGs in 2021, a dramatic increase from five SDGs in 2017. The increased participation of female researchers may help bring about new approaches to achieving SDGs, such as adding gender considerations (Handelsman et al. 2005; Herbert et al. 2022). However, Fig. 5b warns us that women are still severely under-represented in 2021 for certain SDGS, such as SDG13, Climate Action, where gender impacts and promotion of gender equality are especially lacking and needed (Elsevier 2020; Herbert et al. 2022).

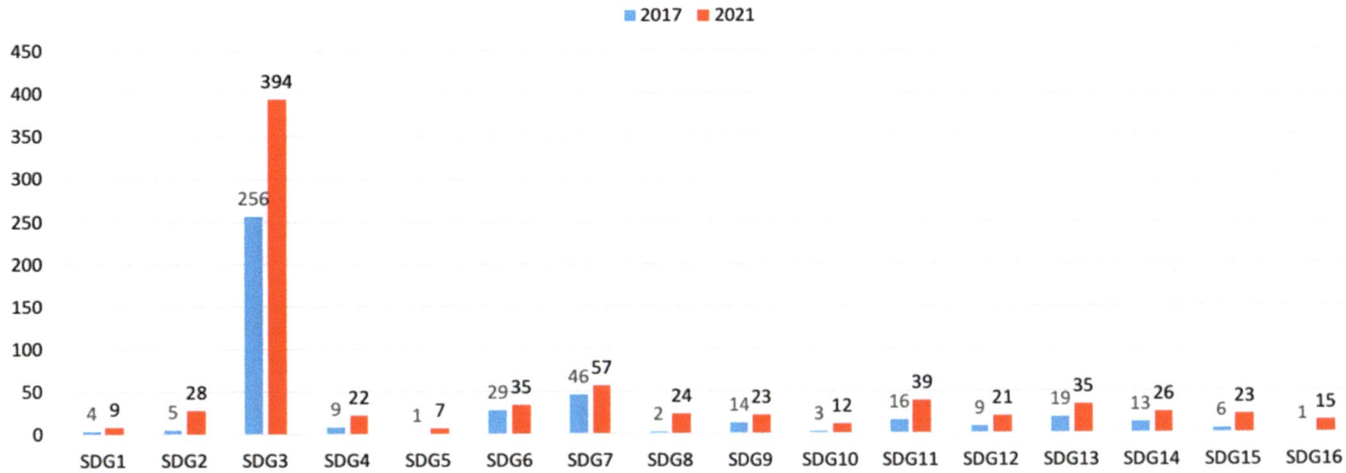

Fig. 3 Number of HU faculty who published articles related to each of the 16 SDGs in 2017 and 2021

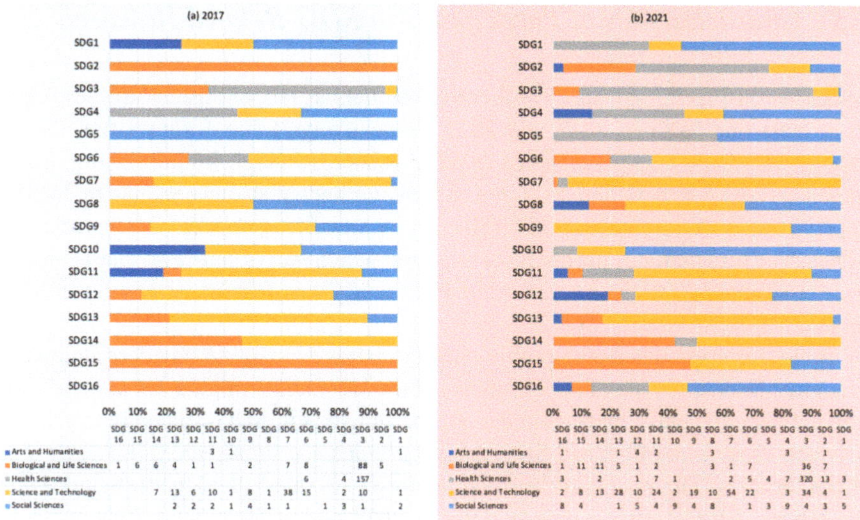

Fig. 4 **a**, **b** Disciplines of HU faculty who published articles related to the 16 SDGs in 2017 and 2021

Fig. 5 **a**, **b** Share of HU women faculty who published articles related to the 16 SDGs in 2017 and 2021

4.3 Characteristics of Participants in SDG-Related Research

As shown in Fig. 1, 361 and 558 faculty members at HU published articles related to at least one SDG in 2017 and 2021, respectively. Table 1 shows that the HU faculty members who participated in SDG-related research by publishing articles increased between 2017 and 2021, not only in number but also in percentage. In 2021, approximately three in ten (29.32%) HU faculty members published articles related to SDGs. The remainder of the analyses explored which individual-level characteristics were associated with the likelihood of HU faculty members participating in SDG-related research.

Table 2 reports the demographic characteristics (discipline, gender, nationality, academic rank, and type of appointment) of the HU faculty population and the variation between the participants and the non-participants in SDG-related research (i.e., between those who published SDG-related articles and those who did not) by these characteristics. The results of bivariate analyses revealed that Health Sciences had the highest participation rate at 37.09% in 2017 and at 50.31% in 2021. Arts and Humanities had the lowest participation rate at 2.40% in 2017 and 4.92% in 2021. The differences across the discipline categories were significant at the $p < 0.001$ level in both 2017 and 2021. The women's participation rate increased from 14.18% in 2017 to 22.28% in 2021, while the men's participation rate increased from 22.92% to 31.22% during the same period, with the gender differences still significant at the $p < 0.001$ level in 2021.

This study categorized faculty into two groups based on nationality: Japanese and non-Japanese. The differences in participation rates were not significant between the two groups in either 2017 or 2021. The experiences of international faculty members working at Japanese universities differ depending on their country of origin. For example, Chinese and Korean faculty tend to be more devoted to research than their American and British counterparts, who are often involved in English language teaching (Huang 2018). Unfortunately, the small number of international faculty members in the sample hampered subgroup analysis by nationality, and there were concerns that individuals could be identified, especially in the 2017 data. Therefore, the broad groups of Japanese and non-Japanese individuals were used in this study.

Table 1 SDG-related article publication participation rates in 2017 and 2021

	2017		2021	
	#	%	#	%
Participants	361	21.53	558	29.32
Non-participants	1,316	78.47	1,345	70.68
Total	1,677	100.00	1,903	100.00

Table 2 Demographics and bivariate analyses for 2017 and 2021

	2017								2021							
	Population		Participants		Non-participants				Population		Participants		Non-participants			
	#	%	#	%	#	%	Chi-square		#	%	#	%	#	%	Chi-square	
Discipline[a]																
Arts and humanities	167	9.96	4	2.40	163	97.60	200.72	***	183	9.63	9	4.92	174	95.08	259.1	***
Biological and life sciences	320	19.08	113	35.31	207	64.69			213	11.20	66	30.99	147	69.01		
Health sciences	426	25.40	158	37.09	268	62.91			648	34.09	326	50.31	322	49.69		
Science and technology	455	27.13	76	16.70	379	83.30			524	27.56	124	23.66	400	76.34		
Social sciences	309	18.43	10	3.24	299	96.76			333	17.52	33	9.91	300	90.09		
Gender																
Men	1,409	84.02	323	22.92	1,086	77.08	10.19	**	1,499	78.77	468	31.22	1,031	68.78	12.28	***
Womens	268	15.98	38	14.18	230	85.82			404	21.23	90	22.28	314	77.72		
Nationality																
Japanese	1,587	94.63	339	21.36	1,248	78.64	0.48		1,691	88.86	508	30.04	1,183	69.96	3.79	
Non-Japanese	90	5.37	22	24.44	68	75.56			212	11.14	50	23.58	162	76.42		
Academic rank																
Assistant professor	514	30.65	119	23.15	395	76.85	14.21	**	708	37.20	208	29.38	500	70.62	22.42	***

(continued)

Table 2 (continued)

	2017								2021							
	Population		Participants		Non-participants				Population		Participants		Non-participants			
	#	%	#	%	#	%	Chi-square		#	%	#	%	#	%	Chi-square	
Koushi (Associate professor/ lecturer)	123	7.33	40	32.52	83	67.48			113	5.94	51	45.13	62	54.87		
Associate professor	448	26.71	78	17.41	370	82.59			489	25.70	115	23.52	374	76.48		
Full professor	592	35.30	124	20.95	468	79.05			593	31.16	184	31.03	409	68.97		
Appointment																
Regular	1,522	90.76	329	21.62	1,193	78.38	0.08		1,669	87.70	517	30.98	1,152	69.02	17.93	***
Non-regular	155	9.24	32	20.65	123	79.35			234	12.30	41	17.52	193	82.48		

Notes * $p < 0.05$, ** $p < 0.01$, *** $p < 0.001$
[a] There are two missing observations in 2021

Participation varied according to academic rank. For example, *Koushis*[2] (associate professor/lecturer) had the highest participation rate, while associate professors had the lowest participation rate in both 2017 and 2021. Differences across academic ranks were significant in both years ($p < 0.01$, 2017; $p < 0.001$, 2021). Approximately ten percent of HU faculty members (9.24% in 2017 and 12.30% in 2021) hold non-regular faculty positions, such as specially appointed professorships and work in collaborative research laboratories. Although faculty members in regular and non-regular positions participated in SDG-related research at a similar rate in 2017, the differences became significant at the $p < 0.001$ level in 2021 (30.98% among regular faculty and 17.52% among non-regular faculty). Although not the focus of this study, it may be useful to check the research expectations of faculty in non-regular positions to any changes between 2017 and 2021 in the portion of their research activities that could lead to publications.

Finally, Table 3 provides the results of the multivariate analyses (logistic regressions) of participation in SDG-related research using the demographic characteristics listed in Table 2. The faculty members were clustered within 78 units (e.g., academic departments and research centers) in 2017 and 81 units in 2021. Because the sample design violated the assumption of independent observations, the models reported here were run with clustering adjustments using Stata's *vce* command.

One major finding from the multivariate analysis was that the faculty in Health Sciences were consistently more likely to participate in SDG-related research than the faculty in other disciplines, controlling for the other variables. Although the faculty in Biological and Life Sciences was the only group that did not show a significant difference from the faculty in Health Sciences in 2017, the difference in the likelihood of participation became significant in 2021 (Odds Ratio [OR] = 0.37, $p < 0.001$). The biggest differences were observed between the Health Sciences and Arts and Humanities faculties (OR = 0.02, $p < 0.001$ in 2017 and OR = 0.03, $p < 0.001$ in 2021).

Another striking finding was the persistent gender gap. In both 2017 and 2021, women were less likely than men to participate in SDG-related research, controlling for other variables (OR = 0.53, $p < 0.05$ in 2017 and OR = 0.63, $p < 0.01$ in 2021). Compared with men, women tend to be in less secure, lower-ranked positions (e.g., untenured, non-tenured track) (Homma et al. 2013; Zippel 2017). In addition, women are underrepresented, especially in Science, Technology, Engineering, and Mathematics (STEM) fields (e.g., Branch 2016; Homma et al. 2013; Goulden et al. 2011). HU is no exception. At HU, women were concentrated in lower-ranked positions and the percentage of women varied from field to field (data not shown). It is intriguing to consider the factors that explain the remaining gender gap after taking into account the influence of position characteristics (academic rank and appointment type) and discipline. Considering the literature related to women's underrepresentation in Japanese academia (e.g., Kitanaka 2014; Normile 2001; Osumi 2006), it is necessary to obtain additional data on faculty work environments (e.g.,

[2] The literal meaning of *koushi* is "lecturer." It is an academic rank between Assistant Professor and Associate Professor in Japanese universities.

Table 3 Multivariate logistic regression analysis for SDG-related research participation in 2017 and 2021

	2017 (N = 1,677)					2021 (N = 1,901)				
	OR		95% CI for OR			OR		95% CI for OR		
Discipline[a]										
Arts and humanities	0.02	***	0.01	–	0.09	0.03	***	0.02	–	0.06
Biological and life sciences	0.80		0.38	–	1.71	0.37	***	0.22	–	0.64
Science and technology	0.25	**	0.10	–	0.65	0.23	***	0.14	–	0.4
Social sciences	0.04	***	0.01	–	0.12	0.08	***	0.05	–	0.12
Gender										
Women	0.53	*	0.30	–	0.92	0.63	**	0.44	–	0.88
Nationality										
Japanese	0.31	***	0.20	–	0.49	0.48	*	0.28	–	0.84
Academic rank[b]										
Assistant professor	0.51	**	0.31	–	0.84	0.48	***	0.36	–	0.65
Koushi (associate professor/ lecturer)	0.79		0.44	–	1.40	0.91		0.51	–	1.61
Associate professor	0.75	*	0.58	–	0.98	0.69	***	0.57	–	0.84
Appointment										
Non-regular	0.87		0.56	–	1.35	0.38	***	0.27	–	0.54

Notes OR = Odds Ratio and CI = Confidence Interval
* $p < 0.05$, ** $p < 0.01$, *** $p < 0.001$
N = 1,901 in 2021 due to two missing observations in the discipline variable Adjusted for department clustering
[a] Health Sciences is the omitted reference group
[b] Full professor is the omitted reference group

relationships with supervisors and colleagues, research resources, gender stereotyping, and discrimination) and non-work-related characteristics (e.g., family care responsibilities and work-life integration) to explore more factors in the gender gap.

While the results of the bivariate analysis in Table 2 indicated no significant difference by nationality, the results of the multivariate analyses in Table 3 show that Japanese faculty members were less likely to participate in SDG-related research (OR = 0.31, $p < 0.001$ in 2017 and OR = 0.48, $p < 0.05$ in 2021). This study only provides a cursory analysis of nationality. Future studies should examine the role of nationality using more detailed categories. The number of international faculty members at HU dramatically increased from 90 in 2017 to 212 in 2021 (see Table 2). If the number continues to increase alongside the internationalization of higher education in Japan (Ota 2018; Yonezawa and Yonezawa 2016), a more detailed categorization and comparison, and even an intersectionality analysis, could be possible.

5 Conclusion

The participation of diverse academic researchers is the key to overcoming complex barriers and realizing potential synergies among the SDGs, including synergies between peace and sustainability (Elsevier 2020; Hill et al. 2014; Page 2007, 2017; Purcell et al. 2019; Sharifi et al. 2020). Previous studies using publication data from academic databases (e.g., Web of Science and Scopus) have shown that researchers based in Western countries predominantly produce publications related to SDGs (Elsevier 2020; Meschede 2020; Salvia et al. 2019; Sharifi et al. 2021; Sweileh 2020). Perhaps because of limited information on individual authors, however, these studies could not analyze researcher diversity in terms of other demographic characteristics, such as discipline or gender.

Given the data limitations in previous studies, this study utilized HU's IR data, which matched the faculty's basic demographic data with publication data extracted from academic databases, to analyze the individual characteristics of faculty members who participated in SDG-related research by publishing articles. Research projects do not always lead to publication, and there are different ways for researchers to make an impact other than publishing (e.g., patents, invited talks, and appearances on the Internet) (Chan et al. 2014a, b, 2016). The intention of this exploratory study was to leverage the IR data already compiled by HU to better understand the diversity of participants in SDG-related research.

To summarize the main findings regarding HU: (1) the disciplinary and gender diversity of research participants represented in each SDG was on the rise, and (2) disciplinary and gender imbalances remained regarding the likelihood of participation in SDG-related research. Regarding the first point, along with the growth of the overall HU faculty that published articles related to SDGs, the disciplinary and gender diversity of the participants contributing to each SDG increased between 2017 and 2021. However, women remain underrepresented in research related to certain SDGs where gender considerations may be needed. It is important for universities to focus on the gender composition of research participants through the SDG. Second, combining all the SDGs, there were significant differences in the likelihood of participation in SDG-related research by discipline and gender in 2017 and 2021, even after controlling for other demographic characteristics. To better understand the mechanism of continuing group imbalances and to find ways to strategically engage more researchers from underrepresented groups, future research should consider not only faculty members' demographic characteristics but also other factors, such as their attitudes toward research, the work environment, and non-work-related conditions.

This study is the first step in the author's attempt to use IR data for research-based interventions. Ironically, the results indicate that the IR data used here were insufficient to discuss ways to solve persistent group imbalances. Therefore, I argue that this study's most important implication for HU and other universities promoting SDGs is that further research with more diverse data sources is required to improve the diversity of participants in SDG-related research. Each data source has its advantages and disadvantages. For example, as this study points out, while publication data obtained from major academic databases are useful for analyzing research trends, they provide limited information on individual researchers. In addition, popular academic databases are biased toward certain disciplines and publications written in English (Gusenbauer 2022; Martín-Martín et al. 2018a, b; Vera-Baceta et al. 2019). Comprehensive consideration of findings from studies using a variety of data (e.g., bibliometric data, existing IR data, data from faculty surveys, and interviews) will be necessary to find ways to involve more researchers from underrepresented groups in SDG-related research and achieve synergies among the SDGs.

Acknowledgements I would like to thank my colleagues at the Office of University Strategy for their encouragement and support in conducting this study. I am also grateful to Hirokazu Yanagihara for his advice on statistical analysis and to the anonymous reviewers for their feedback on an earlier draft. This research was supported by the JSPS KAKENHI (Grant-in-Aid for Early-Career Scientists), Grant Number 20K13905.

Appendix A: Discipline Categories

Owing to changes in the personnel management system, the HU Article Dataset had different categories for faculty's academic disciplines in 2017 and 2021. Five categories were used in 2017, and ten categories were used in 2021. Therefore, the ten categories used in 2021 were merged into five categories to match the 2017 categorization.

10 categories in 2021	Merged 5 categories
–Arts and humanities	–Arts and humanities
–Biological and life sciences	–Biological and life sciences
–Health sciences/clinical medicine	–Health sciences
–Chemistry and chemical engineering –Engineering –Mathematics and Informatics –Physics and earth sciences	–Science and technology

(continued)

(continued)

10 categories in 2021	Merged 5 categories
–Education and sports sciences –Law, political science, economics and sociology –Psychology	–Social sciences

References

Branch EH (ed) (2016) Pathways, potholes, and the persistence of women in science: reconsidering the pipeline. Lexington Books

Bozeman B, Gaughan M, Youtie J, Slade CP, Rimes H (2016) Research collaboration experiences, good and bad: dispatches from the front lines. Sci Public Policy 43(2):226–244. https://doi.org/10.1093/scipol/scv035

Chan HF, Frey BS, Gallus J, Torgler B (2014a) Academic honors and performance. Labour Econ 31:188–204. https://doi.org/10.1016/j.labeco.2014.05.005

Chan HF, Frey BS, Gallus J, Schaffner M, Torgler B, Whyte S (2014b) Do the best scholars attract the highest speaking fees? An exploration of internal and external influence. Scientometrics 101(1):793–817. https://doi.org/10.1007/s11192-014-1379-3

Chan HF, Frey BS, Gallus J, Schaffner M, Torgler B, Whyte S (2016) External influence as an indicator of scholarly importance. Cesifo Econ Stud 62(1):170–195. https://doi.org/10.1093/cesifo/ifv010

De Saá-Pérez P, Díaz-Díaz NL, Aguiar-Díaz I, Ballesteros-Rodríguez JL (2015) How diversity contributes to academic research teams performance. R&D Manag 47(2):165–179. https://doi.org/10.1111/radm.12139

DiTomaso N, Post C, Parks-Yancy R (2007) Workforce diversity and inequality: power, status, and numbers. Ann Rev Sociol 33(1):473–501. https://doi.org/10.1146/annurev.soc.33.040406.131805

Elsevier (2020) The power of data to advance the SDGs. https://www.elsevier.com/__data/assets/pdf_file/0004/1058179/Elsevier-SDG-Report-2020.pdf

Freeman RB, Huang W (2015) Collaborating with people like me: ethnic coauthorship within the United States. J Labor Econ 33(S1). https://doi.org/10.1086/678973

Gusenbauer M (2022) Search where you will find most: comparing the disciplinary coverage of 56 bibliographic databases. Scientometrics 127(5):2683–2745. https://doi.org/10.1007/s11192-022-04289-7

Goulden M, Mason MA, Frasch K (2011) Keeping women in the science pipeline. Ann Am Acad Pol Soc Sci 638(1):141–162

Handelsman J, Cantor N, Carnes M, Denton D, Fine E, Grosz B, Hinshaw V, Marrett C, Rosser S, Shalala D, Sheridan J (2005) More women in science. Science 309(5738):1190–1191. https://doi.org/10.1126/science.1113252

Herbert R, Falk-Krzesinski HJ, James K, Plume A (2022) Sustainability through a gender lens: the extent to which research on UN sustainable development goals includes sex and gender consideration. Plos One, 17(10). https://doi.org/10.1371/journal.pone.0275657

Herring C (2009) Does diversity pay?: Race, gender, and the business case for diversity. Am Sociol Rev 74(2):208–224. https://doi.org/10.1177/000312240907400203

Hill LA, Brandeau G, Truelove E, Lineback K (2014) Collective genius: the art and practice of leading innovation. Harvard Business Review Press

Homma MK, Motohashi R, Ohtsubo H (2013) Japan's lagging gender equality. Science 340(6131):428–430. https://doi.org/10.1126/science.340.6131.428-b

Huang F (2018) International faculty at Japanese universities: their demographic characteristics and work roles. Asia Pac Educ Rev 19(2):263–272. https://doi.org/10.1007/s12564-018-9536-7
Hundschell A, Razinskas S, Backmann J, Hoegl M (2021) The effects of diversity on creativity: a literature review and synthesis. Appl Psychol 71(4):1598–1634. https://doi.org/10.1111/apps.12365
Kasuga F (2021) Science as a common language for contribution to sustainability and peace. Sustain Sci 16(4):1229–1231. https://doi.org/10.1007/s11625-021-00972-5
Khalikova VR, Jin M, Chopra SS (2021) Gender in sustainability research: Inclusion, intersectionality, and patterns of knowledge production. J Ind Ecol 25(4):900–912. https://doi.org/10.1111/jiec.13095
Kitanaka C (2014) Harasumento o umidasu daigakukenkyuu no ba no kouzou [The specific characteristics of harassment on university campus]. Gender and Law 11:33–42
Larivière V, Ni C, Gingras Y, Cronin B, Sugimoto CR (2013) Bibliometrics: global gender disparities in science. Nature 504(7479):211–213
Lerback JC, Hanson B, Wooden P (2020) Association between author diversity and acceptance rates and citations in peer-reviewed earth science manuscripts. Earth Space Sci 7(5). https://doi.org/10.1029/2019ea000946
Liu J, Song Y, Yang S (2020) Gender disparities in the field of economics. Scientometrics 125(2):1477–1498. https://doi.org/10.1007/s11192-020-03627-x
Mannix E, Neale MA (2005) What differences make a difference? Psychol Sci Public Interest 6(2):31–55. https://doi.org/10.1111/j.1529-1006.2005.00022.x
Martín-Martín A, Orduna-Malea E, Delgado López-Cózar E (2018a) Coverage of highly-cited documents in google scholar, web of science, and Scopus: a multidisciplinary comparison. Scientometrics 116(3):2175–2188. https://doi.org/10.1007/s11192-018-2820-9
Martín-Martín A, Orduna-Malea E, Thelwall M, Delgado López-Cózar E (2018b) Google scholar, web of science, and Scopus: a systematic comparison of citations in 252 subject categories. J Informet 12(4):1160–1177. https://doi.org/10.1016/j.joi.2018.09.002
McLeod PL, Lobel SA, Cox TH (1996) Ethnic diversity and creativity in small groups. Small Group Res 27(2):248–264. https://doi.org/10.1177/1046496496272003
Meschede C (2020) The Sustainable Development Goals in scientific literature: a bibliometric overview at the meta-level. Sustainability 12(11):4461. https://doi.org/10.3390/su12114461
Network for Education and Research on Peace and Sustainability (n.d) About NERPS. https://nerps.org/about-2/
Nicoson C (2021) Towards climate resilient peace: an intersectional and degrowth approach. Sustain Sci 16(4):1147–1158. https://doi.org/10.1007/s11625-021-00906-1
Normile D (2001) Women faculty battle Japan's Koza system. Science 291(5505):817–818
Osumi N (2006) Woman scientists in Japan: their situation and goals. IUBMB Life (Int Union Biochem Mol Biol: Life) 58(5–6), 273–278
Ota H (2018) Internationalization of higher education: global trends and Japan's challenges. Educ Stud Japan 12:91–105. https://doi.org/10.7571/esjkyoiku.12.91
Page SE (2007) The difference: how the power of diversity creates better groups, firms, schools, and Societies. Princeton University Press
Page SE (2017) The diversity bonus: how great teams pay off in the knowledge economy. Princeton University Press
Phillips KW, Kim-Jun SY, Shim S-H (2011) The value of diversity in organizations: a social psychological perspective. In: De Cremer D, van Dick R, Murnighan JK (eds) Social psychology and organizations. Routledge/Taylor & Francis Group, pp 253–271
Purcell WM, Henriksen H, Spengler JD (2019) Universities as the engine of transformational sustainability toward delivering the sustainable development goals. Int J Sustain High Educ 20(8):1343–1357. https://doi.org/10.1108/ijshe-02-2019-0103
Rivest M, Kashnitsky Y, Bédard-Vallée A, Campbell D, Khayat P, Labrosse I, Pinheiro H, Provençal S, Roberge G, James C (2021) Improving the Scopus and Aurora queries to identify research that

supports the United Nations Sustainable Development Goals (SDGs) 2021 [Data set]. Elsevier. https://doi.org/10.17632/9sxdykm8s4.4

Salvia AL, Leal Filho W, Brandli LL, Griebeler JS (2019) Assessing research trends related to sustainable development goals: local and global issues. J Clean Prod 208:841–849. https://doi.org/10.1016/j.jclepro.2018.09.242

Sharifi A, Simangan D, Kaneko S (2020) The literature landscape on peace-sustainability nexus: a scientometric analysis. Ambio 50(3):661–678. https://doi.org/10.1007/s13280-020-01388-8

Sharifi A, Simangan D, Kaneko S (2021) Three decades of research on climate change and peace: a bibliometrics analysis. Sustain Sci 16(4):1079–1095. https://doi.org/10.1007/s11625-020-00853-3

Sharifi A, Simangan D, Kaneko S (eds) (2023) Integrated approaches to peace and sustainability. Springer. https://doi.org/10.1007/978-981-19-7295-9

Simangan D, Virji H, Hendrix C, Islam M, Kaneko S, Ma Y-S, Mechler R, Pangotra P, Peters K, Sharifi A, Shams SH (2021) A co-designed heuristic guide for investigating the peace-sustainability nexus in the context of global change. Sustain Sci 16(4):1097–1109. https://doi.org/10.1007/s11625-021-00970-7

Sweileh WM (2020) Bibliometric analysis of scientific publications on "Sustainable development goals" with emphasis on "good health and well-being" goal (2015–2019). Glob Health 16(1). https://doi.org/10.1186/s12992-020-00602-2

United Nations (2012) An agenda for peace: preventive diplomacy, peacemaking and peace-keeping (UN Doc. A/47/277-S/24111). https://digitallibrary.un.org/record/144858#record-files-collapse-header

van Knippenberg D, Schippers MC (2007) Work group diversity 8(1):515–541. https://doi.org/10.1146/annurev.psych.58.110405.085546

Vera-Baceta M-A, Thelwall M, Kousha K (2019) Web of science and Scopus language coverage. Scientometrics 121(3):1803–1813. https://doi.org/10.1007/s11192-019-03264-z

Yang Y, Tian TY, Woodruff TK, Jones BF, Uzzi B (2022) Gender-diverse teams produce more novel and higher-impact scientific ideas. Proc Natl Acad Sci 119(36). https://doi.org/10.1073/pnas.2200841119

Yonezawa Y, Yonezawa A (2016) Internationalization of higher education as a response to globalization: Japan's policy challenges since the 1980s. In: Healy E (ed) Creating social cohesion in an interdependent world: experiences of Australia and Japan. Palgrave Macmillan, pp 191–204. https://doi.org/10.1057/9781137520227_11

Zippel K (2017) Women in global science. Advancing academic careers through international collaboration. Stanford University Press.

Interactions Between SDG 14 (Life Below Water) and SDG 16 (Peace, Justice, and Strong Institutions): A Review of Co-benefits, Synergies, Conflicts, and Trade-Offs

Alva Linner, Ayyoob Sharifi, Dahlia Simangan, Letícia Sarmento dos Muchangos, and Remi Chandran

Abstract Understanding the interactions between SDG 14 and SDG 16 is important as the success of ocean governance relies on effective, accountable, and transparent institutions and representative decision-making. At the same time, ocean governance can help establish peace, justice, and strong institutions in the ocean realm. Mapping the interactions between the two goals is crucial to understand the way forward for implementing them. Despite this, the there is limited and fragmented knowledge on the interaction between the two goals. This chapter aims to provide an overview of the current research landscape on the possible interactions between SDG 14 and 16, to understand better the current research gaps and potential avenues for future research. For this purpose, we reviewed the literature and also sought the opinions of researchers and policy makers via an interactive workshop. The main co-benefit

A. Linner
Lund University, Lund, Sweden

Linköping University, Linköping, Sweden

A. Linner
e-mail: alva.linner@liu.se

A. Sharifi (✉) · D. Simangan
The IDEC Institute, Hiroshima University, Hiroshima, Japan
e-mail: sharifi@hiroshima-u.ac.jp

D. Simangan
e-mail: simangan@hiroshima-u.ac.jp

A. Sharifi
School of Architecture and Design, Lebanese American University, Beirut, Lebanon

L. S. dos Muchangos
The University of Tokyo, Tokyo, Japan

R. Chandran
Remote Sensing Technology Center of Japan (RESTEC), Minato City, Japan
e-mail: chandran_remi@restec.or.jp

A. Sharifi et al. (eds.), *Bridging Peace and Sustainability Amidst Global Transformations*, World Sustainability Series,
https://doi.org/10.1007/978-981-99-7572-3_3

highlighted in the literature is that more equitable and inclusive decision-making processes can render more effective ocean and coastal governance by enhancing contextual knowledge and incentivizing compliance. Research on synergistic interactions between SDG 14 and 16 is lacking, although existing research points out that reducing violence and promoting peace decreases the degradation of the ocean ecosystems, and ocean and coastal governance can reduce the risk of violent conflict. Regarding the trade-offs between SDG 14 and 16, it can be concluded that stricter regulations on marine and coastal resources increase the risks of violent conflicts due to heightened competition over resources. Such regulations can also enforce social injustices, where socially a nd economically disadvantaged groups might face repercussions as a side effect of regulations on ocean usage. A key finding of this study is that there is limited research on the conflicts and trade-offs, between SDGs 14 and 16. More context-specific studies are needed to address this issue. For instance, the pursuit of SDG7 (Affordable and Clean Energy) through hydropower projects could undermine biodiversity on both water and land, resulting in challenges to SDG 14 (Life Below Water) and SDG 15 (Life on Land). Such trade-offs and conflicts should be better addressed in future research.

Keywords SDGs · Life below water · Oceans · Peace · Justice · Co-benefits · Trade-offs

1 Introduction

September 2015 marked an end for the Millennium Development Goals (MDGs), adopted in 2000 at the 2000 Millennium Summit in New York. The MDGs were a collective development agenda with the purpose of reducing poverty in the world by the guidance of eight goals. By September 2015 poverty had been drastically reduced, and three and a half of the goals had been achieved. However, the agenda received substantial critique. The achievements of the goals had been severely uneven across the world and were said to be a 'donor agenda' by the global North. Several of the MDGs' goals, targets, and indicators were noncompatible. Also, by the year 2015, the need for urgent climate action had come to the attention of international organizations, and there was a need to establish a collective global agenda on how to sustainably manage the development of the world. Hence, in September 2015, the Sustainable Development Goals (SDGs) were adopted at the United Nations Conference on Sustainable Development in Rio de Janeiro as a part of the 2030 Agenda. The SDGs consist of 17 goals, focusing on the interconnectedness of environmental, social, and economic aspects to achieve sustainable development, something the MDGs were lacking.

These goals aim to "transform our world". The integrated rationale of the goals has compelled decision-makers, planners, and researchers to address the policy coherence among the SDGs (Glass and Newig 2019; Nilsson and Weitz 2019; Tosun and Leininger 2017). There is a need to understand how the goals are interlinked at different levels, where possible conflicts and synergies abound involving a multitude of actors, sectors, and interests (McCollum et al. 2018; Pradhan et al. 2017).

Every sustainable development goal has seen successes and drawbacks in recent years. Since the 2030 Agenda is comprehensive, it is complex to achieve all its goals and subgoals to the same degree. Whereas sometimes promoting a certain goal can produce co-benefits for another, research has also highlighted the possibility that promoting certain goals can lead to a trade-offs. This calls for an examination of the interactions between the different SDGs. In this chapter, the interactions between SDG 14 and SDG 16 will be examined. By conducting a literature review, the possible co-benefits, trade-offs, synergies, and conflicts between SDG 14 and 16 are highlighted, and the policy implications the interactions pose are explored.

Sustainable Development Goal 14 focuses on the oceans, seas, and marine resources and how to manage and conserve them sustainably. Its targets include reducing marine pollution; protecting and restoring ecosystems; reducing ocean acidification; sustainable fishing; conserving coastal and marine areas; ending subsidies contributing to overfishing; increasing the economic benefits from the sustainable use of marine resources; increasing scientific knowledge, research, and technology for ocean health; supporting small scale fishers; and implementing and enforcing international sea law. The ocean is the world's largest ecosystem, and billions of livelihoods depend on the health of the oceans. Therefore, the sustainable development of the oceans is crucial for strengthening human needs and protecting the environment (United Nations 2022).

Promoting peace and strengthening institutions for more equitable societies is the topic of SDG 16. Its targets include reducing violence everywhere; protecting children from abuse, exploitation, trafficking, and violence; promoting the rule of law and ensuring equal access to justice; combatting organized crime and illicit financial and arms flow; substantially reducing corruption and bribery; developing effective, accountable and transparent institutions; ensuring responsive, inclusive and representative decision-making; strengthening the participation in global governance; provide universal legal identity; ensuring public access to information and protect fundamental freedoms; strengthening national institutions to prevent violence and combat terrorism; and promoting and enforcing non-discriminatory laws and policies.

At the time of writing this chapter, the world is experiencing the highest number of violent conflicts since the Second World War, with two billion people living in conflict-affected areas. About 84.2 million people are currently forcibly displaced, and this number is steadily increasing. Even though representative decision-making has increased since 2015 when the goals were implemented, marginalized groups are still severely underrepresented in a vast majority of institutions and arenas for decision-making (United Nations 2022).

Understanding the interactions between SDG 14 and SDG 16 is important as the success of ocean governance relies on effective, accountable, and transparent institutions and representative decision-making. At the same time, ocean governance can help establish peace, justice, and strong institutions in the ocean realm. For example, the goal of reducing Illegal, Unregulated and Unreported (IUU-) fishing, is closely linked to the goal of combatting organized crime, corruption, and bribery. The same interlinkage can be seen between the goal of promoting the Law of the Sea (UNCLOS) and promoting the rule of law in general. Mapping the interactions between the two goals is crucial to understanding how to implement them. Despite this, the research-based knowledge of the two goals' interaction is fragmented. Therefore, this chapter aims to give an overview of the current research landscape on the possible interactions between SDG 14 and 16, to better understand the current research gaps and potential avenues for future research. For this purpose, we first provide an overview of the interactions reproted in the academic literature. This is followed by reporting the main findings of a workshop organized by the Network for Education and Research on Peace and Sustainability (NERPS) to understand better the linkages from the perspective of scientists and policymakers.

2 Materials and Methods

The method used to conduct the literature review was deductive content analysis. Deductive content analysis can be defined as when the structure of the analysis is operationalized before the analysis of the material is conducted, as opposed to inductive content analysis. The content is extracted from the analysis of theories and empirical insights from previous research. A structured matrix is established as a part of the operationalization (Elo and Kyngäs 2008). In the analysis conducted for this paper the categories of the structured matrix were as follows: the geographic focus of the study, methods, specific SDG 14 target, specific SDG 16 target, type of interaction, details of each type of interaction, other relevant SDGs, and mediating factors. The content that would be extracted was based on the theory that there are four different types of interactions that can occur between different SDGs: co-benefits, synergies, trade-offs, and conflicts (Fader et al. 2018; Sharifi 2020; Singh et al. 2018). To understand the context of the articles and interactions they deal with, it was important to also include what other relevant SDGs could have an impact on the interaction and what factors might mediate the outcomes. The inclusion of geographic focus was also of interest to get an overview of where research on SDG 14 and 16 has been conducted, and if some areas of the world are more under-researched than others.

The definitions used for the different types of interactions were drawn from current IPCC frameworks and SDG-related research. Co-benefits were defined as when there is a positive effect on a target as a result of improving policies and measures aimed at another target (Grafakos et al. 2019). Synergies were defined as when interactions between different policies and measures lead to greater benefits for several targets than when they were implemented separately (Grafakos et al. 2019). Trade-offs were defined as when efforts to enhance one target diminish efforts to achieve another (Berry et al. 2015). Lastly, conflicting interactions were defined as when by implementing one target, it is impossible to implement another, meaning that one cannot achieve both at the same time (Landauer et al. 2015).

Table 1 The string used for literature search in the web of science

TS = ((("peace*" OR "violen*" OR "institution*" OR "corrupt*" OR "law*" OR "crime*" OR "justice" OR "war" OR "access to information" OR "freedom*") AND ("marine" OR "sea*" OR "ocean" OR "coast*")) AND ("synerg*" OR "impact*" OR "trade?off*" OR "interact*" OR "*benefit*" OR "*effect*" OR "*link*") AND ("sustainability" OR "sustainable development"))

A first step in the analysis was to create a search string to find articles of interest. Table 1 displays the search string used in the Web of Science, with keywords included such as sustainable development goal, ocean, marine, conflict, institution, law, etcetera. The search string gave a result of roughly 1300 articles in September 2022. The abstract of each article was read to determine its relevance to the aim of this study, and 90 articles were selected to be included in the literature review. The articles were then read thoroughly to determine what type of interaction the authors were describing and its contexts. The data pre-determined by the structured matrix for each article was extracted and included in an Excel sheet.

When all articles were examined, the results of the literature were synthesized to establish which interactions were most and least common.

A workshop was organized by NERPS to further discuss the interactions and explore potential interactions not mentioned in the literature. The participants were from diverse backgrounds, including resource management, economics, peace studies, remote sensing, etc. We first reported the results of the literature review to the participants. Next, we divided them into two groups to discuss interactions between the two SDGs. The groups were later reshuffled to ensure maximum knowledge exchange between the participants.

3 Results

3.1 Overview

By far, the most common interaction between SDG 14 and 16 found in the literature reviewed was co-benefits, as we can see in Fig. 1. Thirty-nine articles highlighted this interaction, seventeen highlighted possible trade-offs, and six pointed at synergies. No article included in the literature underscored conflicts. The most common targets within SDG 14 interacting with targets of SDG 16, as seen in Fig. 2, were to protect and restore ecosystems (14.2), sustainable fishing (14.4), and supporting small-scale fishers (14.b). The most common targets of SDG 16 were to ensure responsive, inclusive, and representative decision-making (16.7), develop effective, accountable, and transparent institutions (16.6), promote the rule of law, and ensure equal access to justice (16.3), and promote and enforce non-discriminatory laws and policies (16.b), as seen in Fig. 3.

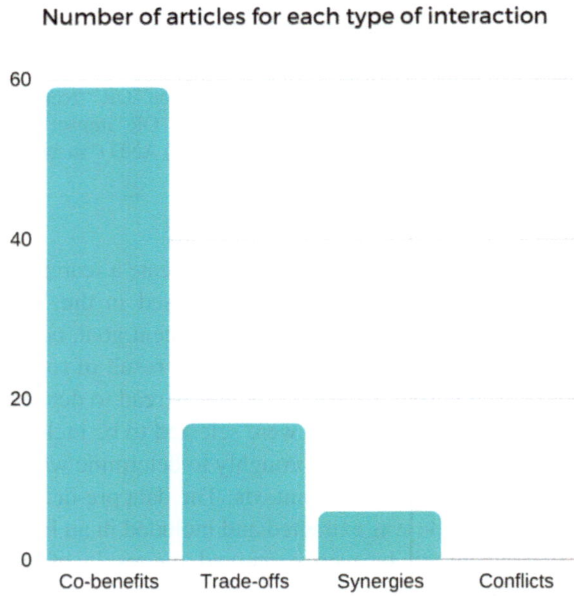

Fig. 1 Number of articles for each type of interaction

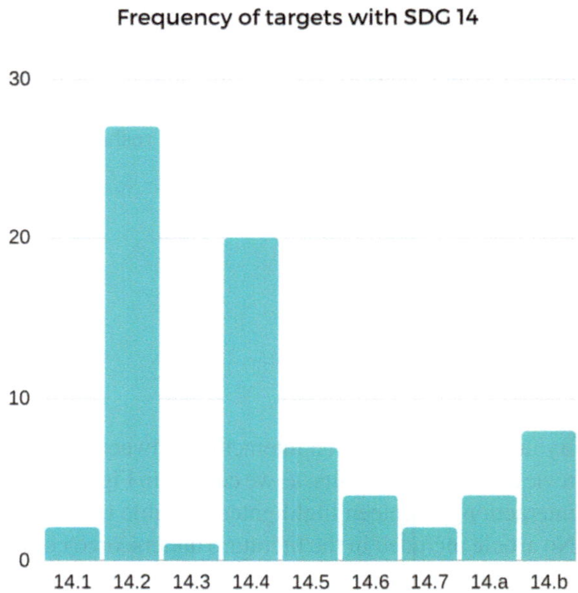

Fig. 2 The number of articles mentioning different targets of SDG 14

From Fig. 4 we see that roughly 50% of all articles included in the literature review were published by four academic journals: *Marine Policy*, *Frontiers of Marine Science, Ocean and Coastal Management*, and *Sustainability*. Figure 5 displays the geographic focuses of the included studies, where a majority (29%) did not include

Fig. 3 The number of articles mentioning different targets of SDG 16

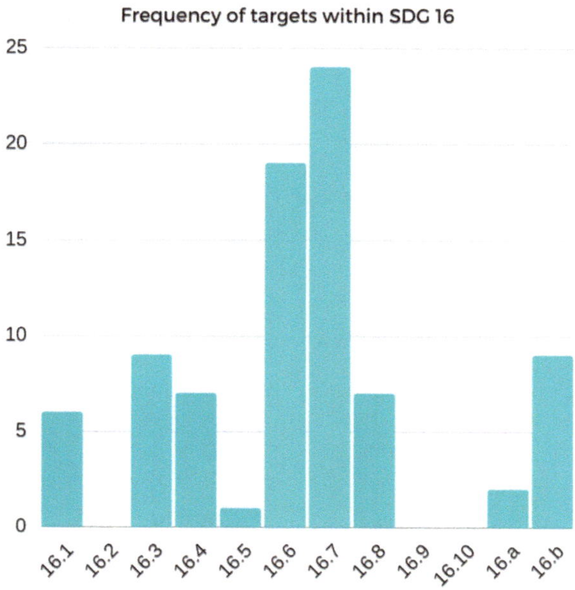

any specific geographic area of study. However, among those who did, Asia and Africa were the most researched continents.

The mediating factors mentioned in the articles were often closely linked to other SDGs. As seen in Fig. 6, the SDGs closely connected to the interactions between SDG 14 and 16 were SDG 8 on decent work and economic growth, SDG 2 on zero hunger, and SDG 17 on partnerships for the goals.

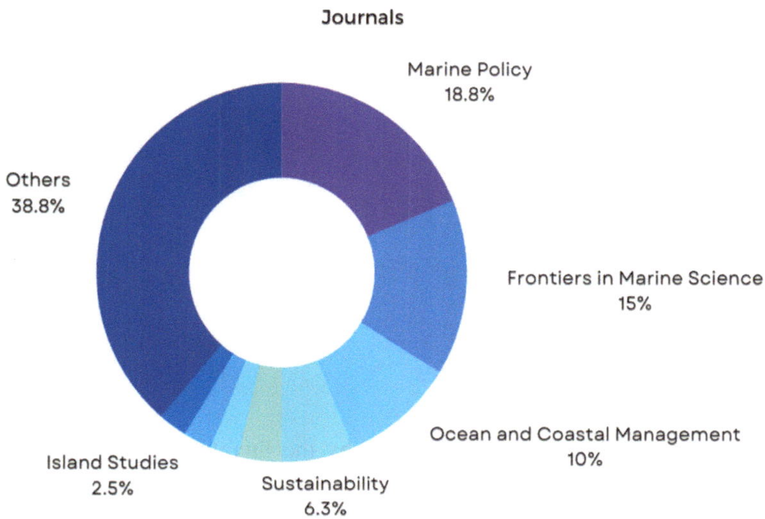

Fig. 4 Major sources of the reviewed articles

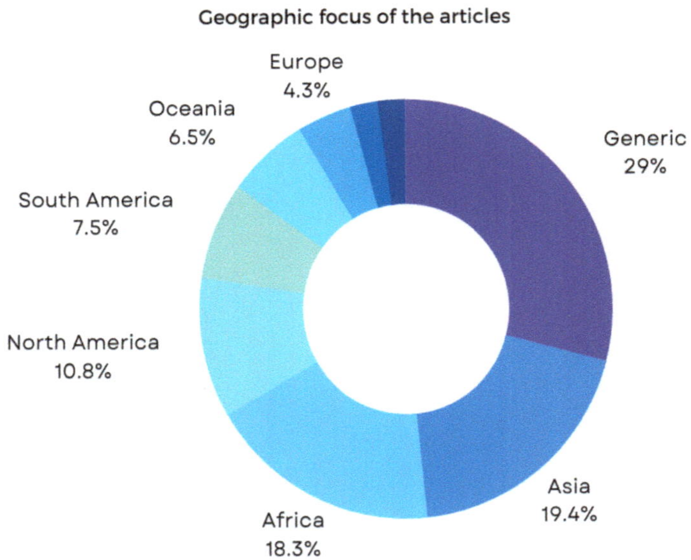

Fig. 5 The geographic focus of the articles

Fig. 6 How often were other SDGs related to the interaction between SDG 14 and 16

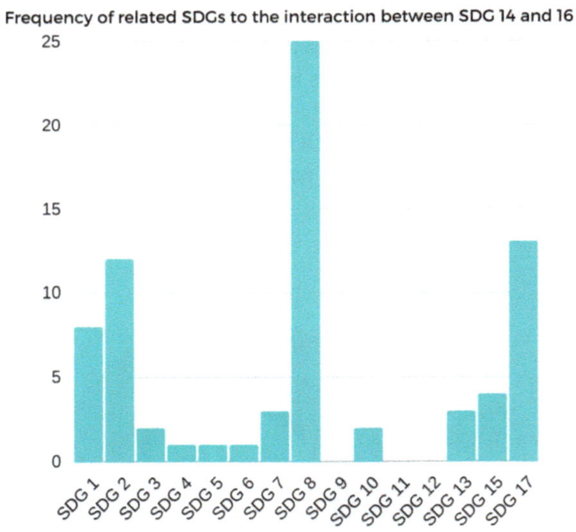

3.2 *Co-benefits and Synergies*

The most common interaction highlighted by the literature regarding co-benefits was between SDG 16.7 (inclusive decision-making) and SDG 14.2 (protecting and restoring ecosystems). Including indigenous peoples and small-scale artisanal fishers in governing institutions and decision-making was highlighted in the literature as beneficial for safeguarding the climate and local ecosystems. Ten different articles pointed to this interaction. Increasing the participation of small-scale fisheries into decision-making processes and monitoring programs is important as their knowledge can contribute to tailoring conservation strategies to the local context. This could increase the effectiveness of the sustainable management of marine and coastal ecosystems (Dias et al. 2020).

By taking a more equitable approach to decision making and governance, the challenges that socio-ecological regime shifts might cause to marine and coastal ecosystems can be detected earlier, and appropriate actions can be taken (Nayak et al. 2016). Dias et al. (2020) also highlight that participation is important to enhance small-scale fisheries' commitment and compliance to regulations, as well as their social justice and legitimacy.

Incorporating indigenous knowledge into decision-making processes and legal framework is also beneficial for conserving the oceans (Tilot et al. 2021). This relationship is because indigenous communities often have insights and knowledge about the local ecosystems, which might be unknown to other groups in society, as they have lived alongside these ecosystems for generations. Indigenous people might have insight into managing the local ecosystems more sustainably. For example, the inclusion of indigenous populations into Arctic governance is beneficial for mitigating the effects of climate change in the region and protecting the local ecosystems through their traditional livelihood strategies (Arruda and Krutkowski 2017). The authors argue that expansionism into resource-rich areas is a form of neo-colonialism as it excludes minority groups, who historically are the primary stakeholders. The Arctic Council needs to reform to include indigenous populations with a participant status to not further exacerbate negative environmental impacts and ensure social equity (Arruda and Krutkowski 2017).

Another co-beneficial interaction highlighted in the literature was between SDG 16.8 (strengthen participation in global governance) towards SDG 14. Strengthening the participation of Small Island and Developing States (SIDS) in global governance institutions enhances experience sharing, and in turn, increases knowledge and appropriate responses to preserve oceans. Six articles highlighted this interaction. This is because the Small Island States often have experience in reacting and pro-acting to climate change. By strengthening the participation of SIDS in institutions of global governance, their experiences can be shared, and in turn, increase the scientific knowledge and appropriate responses to preserve oceans (Poti et al. 2022). Efforts to increase ocean sustainability benefit substantially from increased international diplomacy and strengthened participation of SIDS in global governance. A lack of inclusive international action may stall or even prevent sustainable ocean development in small island states (Singh et al. 2021b).

SDG 16.7 (inclusive decision-making) is also highlighted as beneficial for combatting Illegal, Unreported, and Unregulated fishing (SDG 14.4) in five articles. Participatory decision-making on IUU-fishing can contribute to better compliance with the regulations. More participatory decision-making when it comes to combating IUU-fishing can contribute to better compliance with the regulations by small-scale fisheries, which lowers the prevalence of IUU-fishing (Luomba et al. 2016). One key driver for non-compliance is poverty. Through participatory processes, educating small-scale fishers about the environmental damages that can result from engaging in illegal fishing activities becomes possible. Also, it will be possible to bring the fishers' needs can be brought into the light (Luomba et al. 2016).

A more equitable and participatory decision-making process also ensures that fishers acquire a sense of ownership in fisheries management institutions, which also is beneficial for fishers' compliance with the regulations. When the fisher's knowledge and experience is disregarded in ocean management, it often leads to resentment, which can lead to non-compliance and avoidance of the regulations set in place out of spite (Hart 2021). The increased compliance will ensure sustainable fishing activities (Hart 2021). Using participatory and decentralized processes is therefore beneficial to sustainably manage harvesting and overfishing (González Laxe et al. 2018). González Laxe et al. (2018) go as far as to argue that it is not possible to ensure sustainable fisheries or protection of the environment if the social and economic related issues are not considered. As highlighted above, non-compliance to IUU-fishing regulations from small-scale fishers often is driven by untended socio-economical needs.

There is also a co-beneficial relationship between reducing corruption and bribery (SDG 16.4), and in turn, combatting organized crime (SDG 16.5) and lowering the occurrence of IUU-fishing (SDG 14.4). By accumulating information on how organized crime is operated, including IUU-fishing, governments can better formulate policies to combat organized crime. Eliciting information from fisheries officers provides insights into how government institutions can build a knowledge base on IUU-fishing activity (Donlan et al. 2020). For example, a lot of the IUU-fishing in Somali waters are by foreign vessels which are often given "licenses" to fish by local authorities, leaders, and warlords. This results in vast overfishing, as these licenses allow the foreign vessels to conduct illegal, unreported, and unregulated fishing. By reducing corruption and bribery, and combating organized crime, IUU-fishing can be reduced, lowering the currently high overfishing level (Glaser et al. 2019).

Lastly, there is a co-beneficial relationship between promoting the Law of the Sea (SDG 14.c) and incorporating the agreement into national laws, which promotes rules of law in general (SDG 16.3). Promoting the Law of the Sea encourages states to incorporate the agreement into national laws, which generally promotes rule of law. Also, the law includes agreements concerning armed and military conflict at sea, which can reduce the use of violence (Zou 2021). For example, by promoting sustainability aspects to be included in World Trade Organization (WTO) law, conservation and sustainable use of the oceans can be enhanced through the rule of law (Chong, 2018).

Synergy interactions between SDG 16 and 14 were not as commonly presented in the literature as co-benefits. Only six articles mentioned this interaction. Each synergy interaction only appeared once in the literature included in this study. However, some interesting interactions were highlighted in the articles. Therefore, some synergies between SDG 14 and SDG 16 are presented below.

Ameyaw et al. (2021) point to a synergy between enhancing governance and management of fisheries in developing countries (16.6), and less overfishing (14.4).

In turn, less overfishing lowers the risk of violent conflict over marine resources (16.1). This, in turn, is beneficial for protecting the environment (14.2), as the conflict worsens environmental degradation. The authors use the example of fisheries in Ghana, which many people rely on for food, livelihoods, employment, income generation, and poverty reduction. However, the fisheries face severe overfishing, which may incite conflicts because of competition over resources. If the ocean and fisheries governance in these communities can be enhanced, the threat of overfishing can be decreased. This could also help lower the risk for conflict. Conflicts lead to increased levels of resource competition and degradation and are also a threat to human security. To promote long-term sustainability and stability in the country, ensuring adequate fisheries management is vital (Ameyaw et al. 2021).

Jones and Long (2021) argue that there is a synergy between ensuring inclusive and representative decision-making (16.7) and conserving marine and coastal areas (14.5). Marine Protected Areas (MPAs) provide an opportunity for more equitable and representative decision-making. Representative decision-making is also considered to be one of the main aspects deciding whether establishing a MPA will be effective. In their study, Jones and Long find that the main reason for unsuccessful MPAs was poor governance and corruption. Therefore, equitable decision-making and establishing MPAs work in synergy (Jones and Long 2021).

Jackson et al. (2020) argue that there is a cyclical relationship between modern slavery and environmental degradation, meaning that combatting human trafficking and modern slavery (SDG 16.4) will lower the degradation of the ocean and coastal ecosystems (14.2). Also, protecting and managing the ocean and coastal ecosystems lowers the risk of modern slavery to prevail. Environmental damage is both a driver and a result of workers subjected to modern slavery, including coastal ecosystems like mangroves. Jackson et al. (2020) examines the "fish camps" in coastal Bangladesh, where the use of human slavery is frequent. These "fish camps" are located in protected areas, causing deforestation and damage to the mangroves, contravening coastal conservation efforts. The degradation of the environment also provides space for further human trafficking. Therefore, both conserving marine and coastal ecosystems and combatting organized crime, which includes human trafficking, can enhance each other, working in synergy (Jackson et al. 2020).

To summarize, the main co-benefits highlighted in the literature is that more equitable and inclusive decision-making processes can render more effective ocean and coastal governance by enhancing contextual knowledge and incentivizing compliance. Research on synergistic interactions between SDG 14 and 16 is lacking, although existing research points out that reducing violence and promoting peace decreases the degradation of the ocean ecosystems, and ocean and coastal governance can reduce the risk of violent conflict.

3.3 Trade-Offs and Conflicts

Trade-offs between SDG 14 and SDG 16 are also not as commonly researched as co-benefits, although more than synergies. Seventeen articles from the literature highlighted potential trade-offs between SDG 14 and 16. Some common themes from these articles will be presented in this section.

One potential trade-off between SDG 14 and 16 is that regulating coastal activity (SDG 14.2) can result in social injustices and the exclusion of less powerful people (SDG 16.7), like small-scale and artisanal fishers. Only two articles highlighted this interaction, yet it is interesting to point out since it goes against one of the common co-benefits illustrated in the previous section, which maintained that representative decision-making (SDG 16.7) is beneficial for preserving the ecosystems (14.2). In contrast, the articles by Parlee and Weber (2018) and Hart (2021) describe the difficulties that can arise when ecosystem protection is set in place. Parlee and Webers (2018) present the conflict between representativeness and management of the ocean because of conflicting interests in Canada. There is a vast plurality of actors within the management of the Canadian marine environment. Members of representative institutions are not always representative, and accountability and transparency are not inherent outcomes of participatory processes (Parlee and Weber 2018).

Also, as Hart (2021) highlighted, fishers are largely excluded from fisheries management institutions. Institutions specializing in fisheries biology dominate the fishery-management system. This has diminished the power of fishers over their own activities and equitable decision-making (Hart 2021). This potential trade-off illustrates that caution is needed when strengthening coastal and ocean management. It is important to not assume representativeness without incorporating grass-root organizations and minority communities affected by these regulations. Yet, more actors might lead to less effective institutions, which can be problematic for addressing the issue of ocean health where action is urgent. Again, as seen in the previous chapter, more equitable decision-making is crucial for these actions to be effective in the long run.

Another potential trade-off is that as regulations towards marine protections increase (SDG 14.2), so might stress for families struggling with uncertain fishing futures, increasing domestic violence (SDG 16.1) (Coulthard et al. 2020). The authors studied the prevalence of alcoholism and domestic violence in fishing households in Sri Lanka and India, and found it increased when the family's primary income is restricted. Coulthard and co-authors point out that while supporting small-scale fisheries is one of the SDG targets (14.2), there is often a disregard for how ocean regulations might affect those most dependent on it. Also highlighted in this study is that policies that restrict access to marine resources can undermine important coping strategies, in particular, the ability of women to act as independent income earners. In Sri Lanka and India, marine resources enable financial autonomy for women and bargaining power in households experiencing domestic violence. Poorly informed regulating decisions can inadvertently increase women's vulnerability in compromising positions (Coulthard et al. 2020). This trade-off poses the question of whose need for protection is favored in sustainable management. Several unintentional consequences might emerge from certain policy decisions within sustainable development, which only underlines the importance of thoroughly examining the possible interactions.

Locally Managed Marine Areas (LMMAs) provide opportunities to conserve 10% of coastal ecosystems (SDG 14.5). However, there is an apprehension from fishing communities to adopt LMMAs due to the risk of tensions occurring within and between villages (SDG 16.1) due to conflicting interests and competition over power and resources (Lewis-Brown et al. 2021). The implementation of marine protected areas can trigger conflicts within resource use patterns and power struggles, both internally within a community and between MPA managers and community members. Restrictions on the MPAs have adverse consequences in communities by affecting fishing, tourism, and mining activities. It is uncertain whether MPAs, which are designed on the basis of community involvement, are achieving their ecological and socio-economic objectives effectively (Katikiro 2020).

Efforts to combat IUU-fishing through fisheries governance (16.4) have disproportionally affected small-scale fisheries compared to the industrial sector (14.b). The tendency to frame IUU-fishing as a crime increases the growing criminalization of small-scale fishers, since there is no distinction between small-scale and industrial fleets in the regulations. It contributes to delegitimizing small-scale fisheries by framing them as un-governable (Song et al. 2020). Song and co-authors mention that the IUU-fishing conducted by small-scale fishers has less detrimental impact on the environment and fish population. They argue that the focus should be on industrial vessels, as they represent most of the overfishing and habitual degradation of ocean ecosystems. There is a need for more precise language in regulations surrounding IUU-fishing with a clear distinction between small-scale and industrial fisheries so that more contextual measures can be taken to combat IUU-fishing effectively (Song et al. 2020). Okafor-Yarwood et al. (2022) also highlight this potential trade-off in the development of a Blue Economy in the African Union, where sustainable fishing is one of the main focuses. Efforts to combat IUU fishing in these countries' fisheries governance and laws have disproportionally affected small-scale fishers, and not the industrial sector, who often are the most unsustainable and stand for a large portion of IUU- and unsustainable fishing. While the African Union is said to strive to protect traditional livelihoods, the focus on economic growth leads to small-scale fishers often being outcompeted by the industrial sector. When faced with fishing regulations, small-scale and artisanal fishers are put in an even more vulnerable position (Okafor-Yarwood et al. 2022). Without additional context in IUU-fishing regulations, the efforts to lower IUU-fishing and related organized crime are less likely to be effective and increase the vulnerability of people in already compromised positions.

Mackay et al. (2020) also focus on a potential trade-offs related to combatting organized crime (16.4) and ending IUU-fishing (14.4). The authors argue that there is a narrative in the literature that associates IUU-fishing to other organized crime, like human trafficking and arms smuggling. In their study conducted in Southeast Asia, however, they did not find any empirical evidence linking IUU-fishing to other types of organized crime. Mackay and co-authors (2020) argue that conflating different types of organized crime with IUU-fishing hinders combatting it effectively and might criminalize victims of modern slavery and forced labor within the IUU-fisheries if putting them in the same context as drugs trafficking, for example .

There is no clear case of conflicting interactions between SDG 14 and 16 in the literature reviewed. This suggests one of two possibilities: either there is a significant research gap in the field, or no conflicting interactions exist between SDG 14 and SDG 16. More research is needed to clarify this issue.

To summarize the trade-offs between SDG 14 and 16, it can be concluded that stricter regulations on marine and coastal resources increase the risks of violent conflicts due to heightened competition over resources and decision-making power. They can also enforce social injustices, where socially and economically disadvantaged groups might face repercussions as a side effect of regulations on ocean usage.

3.4 *Insights from the Science-Policy Workshop*

The workshop on "Evaluating the SDGs and their interactions using spatial data and social indicators" was co-organized by the Network for Education and Research on Peace and Sustainability (NERPS) and the Remote Sensing Technology Center of Japan (RESTEC) on November 2nd 2022 in Tokyo, Japan (Fig. 7). The main objective was to further elaborate on the findings of the literature review and explore potentially missing issues and ideas. The main discussions are summarized in terms of 'alignment with the literature review findings', 'power imbalance aspects', 'cross-cutting perspectives and research questions', 'further research into tradeoffs between the SDGs', 'Coordination and collaboration between research/education and policy/practice' and 'contextualizing SDGs'. These are briefly explained in the remainder of this section.

3.4.1 Alignment with the Literature Review Findings

Overall, participants agreed on the issues highlighted by the literature review. Considering the balance between protecting the marine environment and reducing the human crisis within coastal communities, participants pointed out the potential to exacerbate human rights issues by enforcing sustainability frameworks, such as limiting access to food and water resources and negatively impacting the local communities' development. On the other hand, infrastructure built to protect humans from disasters can interfere with marine ecosystems and activities such as fishing and sightseeing.

Concerning ocean governance, fishing communities often experience social and environmental injustices from institutional aspects such as the open markets and de-regulation frameworks. In addition, locations in which resources like oil and gas reserves are found have seen maritime border issues and conflicts among neighboring

Fig. 7 Opening session of the workshop on "evaluating the SDGs and their interactions using spatial data and social indicators"

countries, at times fuelled by the intervention of external and powerful actors—governments and corporations, which in turn see local communities affected and displaced.

Participants highlighted the need for commitment to genuine inclusive decision-making processes in which the local knowledge of communities is harnessed, and law enforcement is a collaborative endeavor supported by the local communities. Moreover, acknowledging the importance and uncertain nature of international cooperation is paramount. Considering the role of education as a bridging sector that can help objectively address the threats while realizing the sustainability framework promises is also important. When developing climate scenarios, this should be paid attention to by targeting different groups and adding local context research focusing on the SDGs' interactions.

3.4.2 Power Imbalance Aspects

The lopsided demand and supply of marine resources between the economically wealthy countries, which hold high consumption levels, and the less financially rich, which are the central source locations, is a driver of unsustainable practices such as overfishing, illegal dumping, and the resulting pollution of the marine environments, from other over-consumption related activities. Thus, two interconnected aspects were highlighted, how fishing practices are warming the seas and the need to explore the catalyst nature of SDG 12 (climate action) further.

Sovereignty matters were also raised, particularly concerning access to data. Countries with high-tech apparatus and real-time information might leverage those advancements to influence local decision-making for their sole benefit while neglecting sustainability issues. Connected with those above, the absence of militarization mentioned in the SDGs was pointed out; it is an aspect that can play an essential role in ocean governance. As such, multilateralism was mentioned as key, combined with the integration of SDG 9 (Industry, innovation and infrastructure), to support the enhancement and transfer of technologies among countries.

3.4.3 Cross-Cutting Perspectives and Research Questions

The key questions and gaps emerged from the discussions are as follows:

Understanding the oceans' role (and presence) is not a given. There are people unfamiliar with the ocean—how to raise their awareness and interest? Can (offline and online) interactive tools and gamification strategies help raise awareness? Oceanography as a broad field can complicate but also offer answers to the efforts to determine the physical boundaries of marine environments and narrow down the required interventions.

Furthermore, the importance of the ocean beyond food security, incorporating climate issues like decarbonization and carbon sequestration, and the link with peace issues needs investigation.

A missing point on SDG 16 is nuclear issues. What is the ruling/international convention on the attack on nuclear facilities (nearby water bodies and dense urban areas)? Why was nuclear energy neglected in environmental discussions? Can Stockholm 50+ offer a platform to include those issues comprehensively?

The academic community contribution can be two-fold. First, providing data compilation/platforms combined with data interpreters or communications that will connect the data users and data generators. Second (and least explored), introducing an in-deep exploration of the notion of uncertainty and concentrating on the unknowns.

3.4.4 Further Research into Tradeoffs Between the SDGs

The participants noted the value of understanding the co-benefits between the SDGs, but they recognized that tradeoffs/conflicts that may potentially arise from the simultaneous implementation of the SDGs remain under researched. For instance, the pursuit of SDG7 (Affordable and Clean Energy) through hydropower projects could undermine biodiversity on both water and land, resulting in challenges to SDG 14 (Life Below Water) and SDG 15 (Life on Land). It may also cause horizontal conflicts within local communities affected by these hydropower projects or vertical conflicts between the local communities and the government or the business sector. If left unmanaged, these conflicts could hinder the relevant targets of SDG 10 (Reduced Inequalities), especially in reducing income inequalities and socio-economic discrimination.

Relatedly, the participants suggest further research into active conflicts (e.g., the ongoing war in Ukraine) where social, political, economic, and environmental conditions could drastically change. Such dynamics must be closely observed and monitored over time to better understand both the immediate and lasting consequences of armed conflict on SDG implementation. Some participants underscored the value of cost-benefit analysis or benefit estimation as a methodology for assessing the environmental and systemic consequences, including both direct and indirect costs/ benefits, while considering the level of uncertainty and complexity of active conflicts.

3.4.5 Coordination and Collaboration Between Research/Education and Policy/Practice

Crucial to the advancement of knowledge about the interactions between the SDGs is the consistent coordination between actors involved in research/education and policy/ practice. For instance, the participants discuss the relationship between government offices and universities, specifically how governments could influence university-led research and education through policy directives and research funding. It was noted during the discussions that this top-down relationship could hinder research activities on important issues that are not prioritized by the government. For example, despite various ecological threats, many policymakers still prioritize hard security issues, such as terrorism and militarization, over ecological security issues that must be resolved in order to achieve the SDGs. A bottom-up process in which universities or educational institutions can also influence or inform policy-making is imperative for providing a comprehensive view of the co-benefits and trade-offs between the SDGs. However, such a process would be challenging unless trust between governments and universities is forged, as the participants duly recognized.

There is also a dissonance between governments and universities when it comes to the language of the SDGs. Researchers, for instance, might be working on a specific issue related to the SDGs without explicit reference to the terms used in the SDG vocabulary. In this case, relevant research activities are not sufficiently captured by policymakers, let alone reviews or studies that aim to collect or synthesize knowledge on the SDGs. Addressing this issue requires open communication based on mutual trust between researchers/educators and policymakers/practitioners. This open communication could be enhanced by ensuring the accessibility of research findings (e.g., translating technical data to political contexts) and the consistency of policy guidelines. The participants agree that open and free access to data is the foundation for open communication and collaboration.

3.4.6 Contextualizing SDGs

In addressing the knowledge gaps of the SDGs, the participants are reminded of the differences in country-level implementation due to varied levels of development and working mechanisms. Developing countries are confronted with more challenges when it comes to meeting the specific targets of the SDGs, and national as well as subnational governments may adopt the targets based on what is workable or acceptable across administrative levels. While this lack of uniformity in terms of implementation creates an incomplete picture of the SDGs' progress, the participants maintained that such differences must be considered when collecting, systematizing, and interpreting SDG indicators. Similarly, institutional barriers and socio-cultural contexts must be integrated into the assessment of SDGs implementation. Doing so mitigates the risk of providing solutions that may exacerbate or unintentionally create problems that could eventually undermine the SDGs.

4 Final Remarks

In this chapter, we endeavored to shed more light on the interactions between SDGs 14 and 16. We found that the research field is fragmented and lacks consensus in several aspects. The most common interaction is the co-benefit between SDG 14.2 and 16.7. This is related to strengthening institutions and enhancing participatory decision-making, especially including small-scale and indigenous fishers. The existing research on synergies between SDG 14 and 16 indicates that reducing violence and promoting peace decreases the degradation of the ocean ecosystems, and ocean and coastal governance can reduce the risk for violent conflict. However, the restricted use of ocean and coastal resources can also lead to increased conflict, and enforce social injustices, which is pointed out by the literature as a potential trade-off between the two SDGs.

There is a clear connection between the interactions between SDG 14 and 16 to SDG 2 on food security, and SDG on economic growth and decent work. This implies that there is a need to include different aspects of human security (e.g., food and livelihood) when implementing policies surrounding the ocean realm. If food security and precariousness are overlooked, there is a risk that the efforts to achieve SDG 14 and 16 increase vulnerability among people already struggling with poverty and uncertain futures.

There is also a connection between the interaction of SDG 14 and 16 to SDG 17 (Partnerships for the goals), where trade and global cooperation to reach the sustainable development goals are included. The literature reviewed made it clear that without global cooperation, reaching SDG 14 and 16 is very difficult, since their topics are globalized issues. Much of the existing research has covered the implications of strengthening institutions and enhancing democratic processes.

A key finding of this study is that there is limited research on the conflicts and trade-offs, between SDGs 14 and 16. More context-specific studies are needed to address this issue.

References

Ameyaw GA, Tsamenyi M, McIlgorm A, Aheto DW (2021) Challenges in the management of small-scale marine fisheries conflicts in Ghana. Ocean Coast Manag 211:105791. https://doi.org/10.1016/j.ocecoaman.2021.105791

Arruda GM, Krutkowski S (2017) Social impacts of climate change and resource development in the Arctic: implications for Arctic governance. J. Enterp Commun People Places Glob Econ 11:277–288. https://doi.org/10.1108/JEC-08-2015-0040

Berry PM, Brown S, Chen M, Kontogianni A, Rowlands O, Simpson G, Skourtos M (2015) Cross-sectoral interactions of adaptation and mitigation measures. Clim Change 128:381–393. https://doi.org/10.1007/s10584-014-1214-0

Borrás MF, Rivera VS (2022) Implementing SDG 14 under principles, ethical values, and environmental and economic justice for small-scale artisanal Fishermen in Costa Rica. Mar Technol Soc J 56:25–30. https://doi.org/10.4031/MTSJ.56.3.14

Branagan M (2022) Nonviolence and sustainability: an indivisible connection. Sustainability 14:6426. https://doi.org/10.3390/su14116426

Chong, W. (2018). International trade, polar environments and sustainable development. *University of Western Australia Law Review, 44*(1), 90–115

Coulthard S, White C, Paranamana N, Sandaruwan KPGL, Manimohan R, Maya R (2020) Tackling alcoholism and domestic violence in fisheries—A new opportunity to improve well-being for the most vulnerable people in global fisheries. Fish Fish 21:223–236. https://doi.org/10.1111/faf.12426

Dias ACE, Cinti A, Parma AM, Seixas CS (2020) Participatory monitoring of small-scale coastal fisheries in South America: use of fishers' knowledge and factors affecting participation. Rev Fish Biol Fish 30:313–333. https://doi.org/10.1007/s11160-020-09602-2

Donlan CJ, Wilcox C, Luque GM, Gelcich S (2020) Estimating illegal fishing from enforcement officers. Sci Rep 10:12478. https://doi.org/10.1038/s41598-020-69311-5

Elo S, Kyngäs H (2008) The qualitative content analysis process. J Adv Nurs 62:107–115. https://doi.org/10.1111/j.1365-2648.2007.04569.x

Fader M, Cranmer C, Lawford R, Engel-Cox J (2018) Toward an understanding of synergies and trade-offs between water, energy, and food SDG targets. Front Environ Sci 6:112. https://doi.org/10.3389/fenvs.2018.00112

Glaser SM, Roberts PM, Hurlburt KJ (2019) Foreign illegal, unreported, and unregulated fishing in Somali waters perpetuates conflict. Front Mar Sci 6:704. https://doi.org/10.3389/fmars.2019.00704

Glass L-M, Newig J (2019) Governance for achieving the sustainable development goals: how important are participation, policy coherence, reflexivity, adaptation and democratic institutions? Earth Syst Gov 2:100031. https://doi.org/10.1016/j.esg.2019.100031

González Laxe F, Bermúdez FM, Palmero FM, Novo-Corti I (2018) Governance of the fishery industry: a new global context. Ocean Coast Manag 153:33–45. https://doi.org/10.1016/j.ocecoaman.2017.12.009

Grafakos S, Trigg K, Landauer M, Chelleri L, Dhakal S (2019) Analytical framework to evaluate the level of integration of climate adaptation and mitigation in cities. Clim Change 154:87–106. https://doi.org/10.1007/s10584-019-02394-w

Hart PJB (2021) Stewards of the sea. Giving power to fishers. Mar. Policy 126:104421. https://doi.org/10.1016/j.marpol.2021.104421

Jackson B, Boyd DS, Ives CD, Decker Sparks JL, Foody GM, Marsh S, Bales K (2020) Remote sensing of fish-processing in the Sundarbans Reserve Forest, Bangladesh: an insight into the modern slavery-environment nexus in the coastal fringe. Marit Stud 19:429–444. https://doi.org/10.1007/s40152-020-00199-7

Jones PJS, Long SD (2021) Analysis and discussion of 28 recent marine protected area governance (MPAG) case studies: challenges of decentralisation in the shadow of hierarchy. Mar Policy 127:104362. https://doi.org/10.1016/j.marpol.2020.104362

Katikiro RE (2020) The role of local institutions in governance of MPAs. Mar Policy 117:103906. https://doi.org/10.1016/j.marpol.2020.103906

Landauer M, Juhola S, Söderholm M (2015) Inter-relationships between adaptation and mitigation: a systematic literature review. Clim Change 131:505–517. https://doi.org/10.1007/s10584-015-1395-1

Lewis-Brown E, Beatty H, Davis K, Rabearisoa A, Ramiaramanana J, Mascia MB, Mills M (2021) The importance of future generations and conflict management in conservation. Conserv Sci Pract 3. https://doi.org/10.1111/csp2.488

Luomba J, Chuenpagdee R, Song A (2016) A bottom-up understanding of illegal, unreported, and unregulated fishing in lake victoria. Sustainability 8:1062. https://doi.org/10.3390/su8101062

Mackay M, Hardesty BD, Wilcox C (2020) The intersection between Illegal fishing, Crimes at Sea, and social well-being. Front Mar Sci 7:589000. https://doi.org/10.3389/fmars.2020.589000

McCollum DL, Echeverri LG, Busch S, Pachauri S, Parkinson S, Rogelj J, Krey V, Minx JC, Nilsson M, Stevance A-S, Riahi K (2018) Connecting the sustainable development goals by their energy inter-linkages. Environ Res Lett 13:033006. https://doi.org/10.1088/1748-9326/aaafe3

Nayak PK, Armitage D, Andrachuk M (2016) Power and politics of social–ecological regime shifts in the Chilika lagoon, India and Tam Giang lagoon, Vietnam. Reg Environ Change 16:325–339. https://doi.org/10.1007/s10113-015-0775-4

Nilsson M, Weitz N (2019) Governing trade-offs and building coherence in policy-making for the 2030 Agenda. Polit Gov 7:254–263. https://doi.org/10.17645/pag.v7i4.2229

Okafor-Yarwood I, Kadagi NI, Belhabib D, Allison EH (2022) Survival of the richest, not the fittest: how attempts to improve governance impact African small-scale marine fisheries. Mar Policy 135:104847. https://doi.org/10.1016/j.marpol.2021.104847

Parlee CE, Wiber MG (2018) Using conflict over risk management in the marine environment to strengthen measures of governance. Ecol Soc 23, art5. https://doi.org/10.5751/ES-10334-230405

Poti M, Hugé J, Shanker K, Koedam N, Dahdouh-Guebas F (2022) Learning from small Islands in the Western Indian Ocean (WIO): a systematic review of responses to environmental change. Ocean Coast Manag 227:106268. https://doi.org/10.1016/j.ocecoaman.2022.106268

Pradhan P, Costa L, Rybski D, Lucht W, Kropp JP (2017) A systematic study of sustainable development goal (SDG) interactions: a systematic study of sdg interactions. Earths Future 5:1169–1179. https://doi.org/10.1002/2017EF000632

Sharifi A (2020) Trade-offs and conflicts between urban climate change mitigation and adaptation measures: a literature review. J Clean Prod 276:122813. https://doi.org/10.1016/j.jclepro.2020.122813

Singh GG, Cisneros-Montemayor AM, Swartz W, Cheung W, Guy JA, Kenny T-A, McOwen CJ, Asch R, Geffert JL, Wabnitz CCC, Sumaila R, Hanich Q, Ota Y (2018) A rapid assessment of co-benefits and trade-offs among sustainable development goals. Mar Policy 93:223–231. https://doi.org/10.1016/j.marpol.2017.05.030

Singh GG, Oduber M, Cisneros-Montemayor AM, Ridderstaat J (2021b) Aiding ocean development planning with SDG relationships in small Island developing states. Nat Sustain 4:573–582. https://doi.org/10.1038/s41893-021-00698-3

Song AM, Scholtens J, Barclay K, Bush SR, Fabinyi M, Adhuri DS, Haughton M (2020) Collateral damage? Small-scale fisheries in the global fight against IUU fishing. Fish Fish 21:831–843. https://doi.org/10.1111/faf.12462

Tilot V, Willaert K, Guilloux B, Chen W, Mulalap CY, Gaulme F, Bambridge T, Peters K, Dahl A (2021) Traditional dimensions of seabed resource management in the context of Deep Sea Mining in the Pacific: learning from the socio-ecological interconnectivity between Island communities and the ocean realm. Front Mar Sci 8:637938. https://doi.org/10.3389/fmars.2021.637938

Tosun J, Leininger J (2017) Governing the interlinkages between the sustainable development goals: approaches to attain policy integration. Glob Chall 1:1700036. https://doi.org/10.1002/gch2.201700036

Zou K (2021) Climate change and fisheries regulation: what we should consider for the future? Sustainability 13:9735. https://doi.org/10.3390/su13179735

Ecological Education in Islamic Religious Learning Based on Creative Imagination

Desfa Yusmaliana, Fifin Fitriana, Gamal Abdul Nasir Zakaria, and Aji Kurbiyanto

Abstract This study aims to analyze the insertion of environmental education into Islamic religious learning with a creative imagination approach. In the context of the environment, Indonesia is currently experiencing an increasingly severe ecological crisis; education practitioners are then in the spotlight in developing a more meaningful education, especially in Islamic religious education. The problem of values on the environment is shown in the practice of arbitrarily exploiting and not caring about nature. So far, ecology education in Islamic education has used a textual approach, not an approach that has a transformative impact, and creative imagination that comes from neuroscience. Whereas early age is an age that has the potential to be stimulated to the maximum. In this case, the Bangka Belitung Nature School (SABB) has started the concept of learning with nature, which is integrated into all subjects, including Islamic education. This research is phenomenological qualitative research involving principals and teachers at SABB. Data collection techniques were carried out by observation and in-depth interviews. The data analysis used interpretative phenomenological analysis with repeated transcript reading, initial noting, emergent themes, and subordinate themes. The results show that environmental education is inserted into relevant topics, especially Islamic education, based on creative imagination. The investigation results show that an educational model that can be applied to bring out the value of love for the environment is ecological education in Islamic learning based on creative imagination.

Keywords Ecology · Islamic education · Creative imagination · Education model

D. Yusmaliana (✉) · F. Fitriana · A. Kurbiyanto
Universitas Muhammadiyah Bangka Belitung, PGSD, KSDA, Pangkalpinang, Indonesia
e-mail: desfa.yusmaliana@unmuhbabel.ac.id

D. Yusmaliana · G. A. N. Zakaria
Universiti Brunei Darussalam, SHBIE, Darussalam, Bandar Seri Begawan, Brunei

A. Sharifi et al. (eds.), *Bridging Peace and Sustainability Amidst Global Transformations*, World Sustainability Series,
https://doi.org/10.1007/978-981-99-7572-3_4

1 Introduction

The Intergovernmental Panel on Climate Change (Delmotte et al. 2021) and Watsiqotul (2018) noted that environmental damage, depletion of natural resources due to exploitation, and increased global warming result from human actions. Damage to the soil, water, and air, deforestation and forest degradation, and even forest fires that often occur, resulting in the depletion and destruction of biodiversity, including sea level rise and the spread of various types of diseases are various forms of problems called environmental crises (Syamsudin 2017). Whereas in Islamic teachings, there are many hadiths and at least 800 verses regarding the obligation to manage nature and protect it (Nurhayati et al. 2018) because they will be held accountable so that humans are not entitled to lead and manage nature arbitrarily. For this reason, it requires redrawing the relationship between humans and nature.

Efforts to develop environmental awareness are a process that must be initiated for the survival of the next generation so as not to be threatened by irresponsible human behavior. Ecology education in religious learning aims to introduce various values and explain concepts to develop the attitudes and skills needed to understand and appreciate the interrelationships between humans, culture, and the biophysical environment with an early creative imagination approach.

Imagination, in a broad view, is often considered a futile work, even though, in its application, imagination is included in the realm of creative thinking (Yusmaliana et al. 2020). Even imagination, part of the subconscious mind, contributes enormous power in making decisions or actions. Dijksterhuis (2004) experiment found that unconscious thinkers made better decisions, resulting in clearer, more polarized, and more integrated memory representations. For this reason, educators must understand that humans will be more easily given information and directed through their subconscious activities by touching their emotions and maximizing their imagination.

Educational institutions then become a benchmark in developing students' minds or thinking abilities. Even in the face of global competition and providing solutions to various existing problems. For this reason, the concept of education, as Ahmad Tafsir expresses, is guidance to someone who must be given optimally (Basri 2013). Even in the 21st century, creative thinking, problem-solving, innovation, and communication are required. Therefore, the curriculum will continuously develop in learning, including Islamic religious education. Yusmaliana and Widodo (2019) revealed that the reconstruction of Islamic education in the era of disruption is to develop character education with a learning approach using various technologies to create a learning atmosphere that inspires, enlightens, and builds Islamic values. In its development, there have been many media and applications that integrate nature education into subjects, such as what was done by Winda Purnama Sari and Aprilliandari (2021). They developed the BioEnviroScience App (BES App) for learning environmental science materials, especially in the sub-environmental change in the Province of the Bangka Belitung Islands. Likewise, Indra and Fitria (2021) developed an educational science game media assisted by the Appsgeyser application to improve the environmental care character of elementary school-aged children. However, the integration of natural education with Islamic education is minimal, especially using creative imagination. Imagination is still often regarded as mere fantasy and does not contribute anything to the cognitive development of students (Yusmaliana and Suyadi 2019b).

For educators, cultivating an understanding of ecology is a much more challenging task than imparting knowledge on ecological issues—something that still tends to dominate some ecologically oriented educational programs. Educating for ecological understanding requires focusing on engagement during teaching and learning (Asroni 2021). To that end, emotional and imaginative engagement with knowledge, the world, technology, and the natural and cultural contexts in which students live and learn can support this decision-making and action. For example, one must wisely refuse, reduce, reuse, and recycle.

In the science that studies the brain, it is known that six components in the brain are closely related to a person's character, including the character of loving the environment, namely the 'cortex prefrontalis, system limbic, ganglia basalis, gyrus cingulatus, lobus temporalis, and cerebellum' (Pasiak 2010). These six parts are interrelated and provide instructions for humans to do or not to do something (Suyadi et al. 2021). Therefore, the task of Islamic religious education is to direct, provide knowledge, and give examples to stimulate someone, in this case, students, to maximize their love for their environment. Therefore, Islamic values that teach a lot about the importance of being grateful and preserving the environment must be integrated into fun learning.

As the researcher's statement at the beginning that cultivating an understanding of ecology is far more challenging than providing knowledge about ecological issues is the same as the difficulty in providing an understanding of the application of Islamic values to students' real lives. The challenge is in helping students develop an appreciation for the natural world and an understanding of the importance of environmental conservation. This requires not only teaching students about the impacts of human activities on the environment but also helping them to see the value in preserving and protecting natural habitats and species. By instilling a sense of wonder and curiosity in their students, educators can inspire them to become stewards of the environment and make a positive impact on the world around them (Barrett 2001). Unfortunately, given the importance of the study of ecological education in Islamic education and the creative imagination approach, there are still many gaps and limitations. For this reason, a new approach is needed in Islamic education toward applying Islamic values, especially related to existing environmental issues, such as a neuroscience-based creative imagination approach in the learning process.

Overall, teaching ecology is a challenging but rewarding task for educators. By helping students develop an understanding and appreciation for the natural world, educators can play a crucial role in shaping the next generation of environmental stewards and promoting sustainable practices. Thus, this study is intended to fill this gap. Islamic Religious Education can direct, provide knowledge and provide examples to be able to stimulate someone, in this case, students, to be able to maximize love for their environment. Indeed, Islamic values should be integrated into enjoyable learning, which teaches a lot about the importance of being grateful and preserving the environment.

2 Literature Review

2.1 Negative Behavior or Destruction of Nature

Destruction of nature is a negative behavior that is repeated throughout history or a period of a person's life. Both intentionally and unintentionally continue to be carried out and impact the environment. Sulistyo called this situation an environmental crisis where humans interacting with the environment are the main subjects determining whether or not an environment is in crisis. It is because humans have two roles, namely as custodians or polluters or destroyers (Sulistyo 2018). As custodians, humans can carry out various activities to protect nature, such as processing existing waste, behaving environmentally friendly by reducing the use of plastic materials that are difficult to decompose after their use, increasing green plants around, and so on. On the other hand, as polluters, humans will cause natural crises and imbalances.

In general, factors of natural damage can indeed be categorized as the result of both natural events and human activities. However, the main cause of the natural crisis is the wedge of the human moral crisis. As stated in QS Ar Ruum verse 41: "*Corruption has appeared on land and at sea due to human actions, so that Allah may feel for them some of the consequences of their actions so that they return (to the right path).*" Seyyed Hossein Nasr in (Abd. Aziz 2014) reveals that one of the symbols of God is nature so that both the mercy and the evil of God will be delivered one of them through this symbol.

Many problems related to the environment stem from a lack of human care and responsibility, both in protecting and preserving the environment. For example, plastic waste is increasingly scattered in the surrounding environment, even though the waste is tough to decompose, causing damage to nature and the environment (Baro'ah and Qonita 2020).

The moral crisis that occurs increasingly illustrates that education has failed to achieve the fundamental learning objectives. The environment will be increasingly damaged if these problems are not considered and remain unchecked. Due to these conditions, giving an understanding of environmental care to the younger generation, especially in early childhood, is a necessity. Early age is a critical period in character building where they can absorb as much as 80% of information and will apply it in their lives in the future (Nizar 2021).

2.2 Ecological Education in Islamic Religious Learning

The integration of ecological education into Islamic religious learning is of significant importance in today's world. As environmental issues continue to escalate, religious teachings must emphasize the responsibility of humans towards the natural world. By incorporating ecological principles into Islamic education, students can develop a deeper understanding and appreciation of the environment.

Inspiring learning is an essential stimulus that can ignite students' imaginations. One effective approach to environmental education is the use of stories, particularly those related to the environment. Suyadi (2018) reveals that religious, inspirational, and motivational stories can foster a strong work ethic and spirituality, with implications for the progress of a nation.

The Qur'an contains numerous references to the creation of the heavens, earth, and various natural resources. These include wind, water, plants, and animals, which are seen as a mercy from Allah to humans (QS. Al A'raf: 57), a source of wisdom (QS. Shad: 27), and signs of Allah's greatness (Surah Al Baqarah: 164; QS. Al A'raf: 58). Additionally, Kholis and Aulassyahied (2022) have discovered at least thirty hadiths that provide a foundation for ecological maintenance and addressing environmental issues.

The term 'ecology' represents the intersection of science and the relationship between creatures and the environment. This connection, known as ecology (Watsiqotul 2018), encompasses extensive components, both biotic and abiotic. The words 'science' (logos) and 'house' or 'residence' (oikos) in ecology are associated with understanding natural conditions, the existence of organisms, and the interactions within ecosystems (Bashyroh and Mahmud 2021).

In a study on righteousness, Pr (2022) highlights the need to increase ecological piety through simple yet meaningful actions to care for the environment and the Earth. Examples include avoiding the use of plastic and paper, reducing water consumption, sorting waste, practicing mindful consumption, treating other living beings with kindness, utilizing public transportation or carpooling, planting trees, and conserving energy by turning off unnecessary lights.

Building an ecologically conscious religious spirit should be pursued simultaneously and sustainably at all levels of society, particularly among Muslims. Recognizing the universal impact of the ecological crisis, re-understanding the role of humans as caliphs in protecting and managing nature is expected to raise public awareness.

It is agreed that all religions acknowledge the regulation of human life, including the interaction between humans and nature. The relationship between humans and nature is comprehensive, extending to how humans behave towards the environment. In the context of Islam, the Prophet's teachings encompass numerous verses and hadiths that promote the protection and preservation of the environment. Nature, in the Islamic perspective, encompasses everything created by Allah SWT, extending beyond celestial bodies and the Earth to everything in between. Thus, nature is a complex and broad creation (Muhaimin 2015). While humans are allowed to utilize and empower natural resources for prosperity and problem-solving, it is essential to avoid causing harm to nature.

2.3 Creative Imagination of Early Childhood

Some educators and educational philosophers who study brain science are starting to understand that the passion for imagination must be included in the learning process. It is not only quarantined in art subjects (Heath 2008) because human imagination will never go extinct and will continue to produce new creativity (Gunarti 2013). From the perspective of neuroscience, the brain that is maximized to function and develop rapidly from an early age is a source of the formation of imagination and creativity (Pasiak 2007). The neurons in the brain will grow more when someone uses them to the maximum, one of which is creativity, which is the highest point of the thinking process (Sousa 2012) and is the primary key to achieving success (Tyan 2005). The creative activities will emerge from the work process of the intuitive brain after experiencing a process of deep analysis and thought. Intuition is a holistic form that can be superior since people with this ability develop and maintain their intuitive brain as an important role in strategic decision-making. In reality, logical people are often more valued than intuitive people. On the other hand, research (Shalihin et al. 2021) shows that the transmutation from the rational to the intuitive brain through this saturation point is included in Higher Order Thinking (C4-C).

In Islamic education, part of the Prophet's learning model includes learning by fostering imagination, curiosity, and the courage to try (Faaizun 2014). The three teaching processes and the concept of creative imagination taken from various opinions of psychologists and learning experts are combined to achieve learning objectives. Thambu et al. (2021) even revealed that integrating active learning into the teaching and learning of moral education can develop a person's multiple intelligences.

Likewise, much earlier, Widiawati (2019) cited Al-Farabi (1968) in revealing that humans acquire knowledge through three powers: the power of the senses, which capture a material object; the power of imagination, which composes and combines stored depictions into new depictions; and the power of thought, which makes decisions on truth. For this reason, stimulating students' creative imagination through various cognitive tools (Yusmaliana and Suyadi 2019a) is a learning strategy that will then be developed, especially in dealing with the current environmental crisis.

One of the *cognitive tools* that can be used at all levels is through experience. Fan and Xiao (2015) revealed that meaningful learning can be obtained from experiences realized through the surrounding environment such as through play activities. Playing activities can encourage development that arises from the interconnection of imagined situations with the game's rules (Lima 2018; Schousboe and Winther-Lindqvist 2013; Sombrio et al. 2014). It will give a lot of experience accumulation to them, which is a factor in realizing creative imagination (Pelaprat and Cole 2011; Sombrio et al. 2014). However, the realization must, of course, be supported by appropriate learning strategies. Therefore, the development of creative imagination in students must also start from the creative imagination of educators, as (Sousa 2012) reveals that a teacher is not an expert in the field of the brain. Still, it is the teacher who will deal with the working system of the brain of students every day.

3 Method

This research is qualitative with a phenomenological approach (Creswell 2012). The study aims to conduct a phenomenological analysis of the insertion of ecological education based on creative imagination in Islamic schools. The research is conducted at Sekolah Alam Bangka Belitung (SABB). The location of this research was chosen because (1) SABB has a curriculum that teaches explicitly based on Islamic values, and (2) SABB has inserted ecological education into learning.

Research Questions:

1. What are the experiences and perceptions of teachers and students regarding the insertion of ecological education based on creative imagination in Islamic schools, specifically in the context of Sekolah Alam Bangka Belitung (SABB)?
   - This research aims to explore the subjective experiences and perspectives of teachers and students regarding the integration of ecological education and creative imagination in SABB, an Islamic school with a curriculum explicitly based on Islamic values and an embedded focus on ecological education.
2. How does the integration of ecological education and creative imagination in SABB impact students' understanding of environmental concepts, their engagement in sustainable practices, and their overall environmental awareness?
   - This research seeks to investigate the effects of the integration of ecological education and creative imagination on various aspects of students' environmental literacy, including their understanding of environmental concepts, adoption of sustainable practices, and overall environmental awareness.

These research questions aim to seek the experiences, effects, and outcomes of the integration of ecological education and creative imagination in SABB, specifically focusing on teachers' and students' perspectives, understanding of environmental concepts, engagement in sustainable practices, and environmental awareness.

Data Collection: The research methods and data collection techniques, such as observation, in-depth interviews, and interpretative phenomenological analysis, are aligned with these objectives to provide a comprehensive evaluation of the research topic. The data source in this research is the process of teaching and learning on subjects that have inserted Islamic values and ecological education. In addition, data was also collected from teachers who teach subjects with embedded Islamic values, ecological education, and learning methods based on creative imagination.

Data collection techniques included observation and in-depth interviews. Observations were conducted from July 2022 to August 2022, focusing on the involvement of teachers in designing and implementing behavioral interventions for teaching ecology education. Interviews about Islamic education, ecological education, and creative imagination were conducted with all informants using the Focus Group Discussion (FGD) method.

The data analysis used interpretative phenomenological analysis with repeated transcript reading, initial noting, emergent themes, and subordinate themes (Smith et al. 2009). The analysis process involved interpretative phenomenological analysis, which consisted of multiple steps to explore the data collected from observations and in-depth interviews with teachers and students. The analysis aimed to identify emergent themes, establish connections between these themes, and develop an integration model of ecological education based on creative imagination.

First, the transcripts obtained from the interviews and observations were repeatedly read to gain familiarity with the data and critically review the learning materials, designs, and concepts implemented by teachers at SABB. Additionally, relevant references listed in the semester's lesson plan were thoroughly studied.

Next, initial notations were made to record the content of each sub-chapter of learning material for each meeting, including both face-to-face and online interactions. This process involved analyzing the specific content related to Islamic education, ecological education, and creative imagination, as well as identifying any overlaps or intersections between these areas.

During the in-depth investigative interviews with teachers and students, emerging themes were identified. These themes were derived from the discussions held during class sessions and special assignments. They represented key concepts and ideas that emerged from the integration of Islamic education, ecological education, and creative imagination. The subsequent step involved establishing connections across these themes. This included exploring the relationships and interdependencies between the study materials within the fields of Islamic education, ecological education, and creative imagination. The in-depth investigative interviews played a crucial role in uncovering these connections and providing a deeper understanding of how these themes were intertwined.

Finally, the researchers looked for patterns and developed an integration model of ecological education based on creative imagination. This model synthesized the emerging themes and their connections, providing a comprehensive framework for understanding how the integration of Islamic education, ecological education, and creative imagination can be effectively implemented in educational settings.

4 Results and Discussion

The following section presents the results, focusing on the experiences and perceptions of teachers and students regarding the integration of ecological education and creative imagination at Sekolah Alam Bangka Belitung (SABB), as well as its impact on various aspects of students' environmental literacy and engagement.

In response to Research Question 1 ("What are the experiences and perceptions of teachers and students regarding the insertion of ecological education based on creative imagination in SABB?"), the data revealed a consistent theme among both teachers and students regarding the positive impact of ecological education on their understanding of environmental concepts. Teachers expressed that integrating creative imagination in the curriculum enhanced students' engagement and deepened their comprehension of ecological principles. Similarly, students reported a heightened sense of connection with nature and an increased understanding of sustainability through the integration of creative imagination in their learning experiences.

Moechiji (2022) revealed that SABB has not implemented a curriculum based on creative imagination. However, all activities should build students' imagination. At the time of learning, students are directed to five models: conservation, experimentation, survey, project, and learning by wisdom. As the outline and framework contained in the concept of Islamic religious learning are based on neuroscience-based Creative Imagination (Yusmaliana et al. 2022a), namely: finding the exciting side of the material to be taught first to arouse curiosity, using cognitive tools such as stories, puzzles, riddles, jokes or humor, poems, games, dramas, and so on provide a stimulus to students to get conclusions or lessons learned that have occurred, and finally evaluate learning.

One of the essential informants in this research is the principal, who is the director of the application of ecological education and Islamic values. According to him, three competencies are often not maximally applied by teachers in school education, namely: (1) teachers must inspire, (2) teachers must be role models, and (3) teachers must design interesting and fun learning (Syahril 2022). In Islamic religious learning, the Bangka Belitung Natural School (SABB) admits to integrating it into every subject. The same applies to ecological education. However, it has not been explicitly summarized in one complete design.

In an in-depth analysis, SABB inserts ecological education into the curriculum with a design of learning achievement on each meeting theme until it reaches a conclusion, "*Ma Sha Allah*" or meaning that students know that it is the greatness of Allah that everything can happen. Allah has created for the benefit of humans, so it is an obligation for humans to maintain and preserve their natural surroundings.

Research Question 2 ("How does the integration of ecological education and creative imagination in SABB impact students' understanding of environmental concepts, their engagement in sustainable practices, and their overall environmental awareness?") yielded noteworthy findings. The data indicated that students who experienced the integration of ecological education and creative imagination demonstrated a significantly higher level of understanding of environmental concepts compared to students in schools without such integration. Moreover, the students' engagement in sustainable practices, such as waste reduction and conservation efforts, showed a marked improvement, aligning with their enhanced environmental awareness.

The following is an ecological education activity in Islamic religious learning that occurs at the Bangka Belitung Nature School (SABB).

3. The teacher gave directions and examples of some acts of damage to nature. Students were asked to select and stick the appropriate image and imagine if the damage occurred around their living (Fig. 1).
4. The teacher invited students to play "the paths of nature trails" and asked them to reflect on what they had encountered to conclude how beautiful the natural surroundings are and how lucky humans are to be blessed with this beautiful

Fig. 1 Students imagine if their surroundings are damaged

nature. In this part, again, a necessary component of learning is the teacher. Teachers need to know how to stimulate students' learning in the right ways when it comes to learning (Wantini et al. 2022) (Fig. 2).

Although the majority of the results was supported and indicated positive outcomes, there were a few unexpected findings. Some students who initially showed resistance to the integration of creative imagination expressed challenges in applying their newfound understanding of environmental concepts to real-world situations. This unexpected outcome may be attributed to the need for further reinforcement and practical application opportunities in the curriculum, suggesting potential areas for improvement.

Ecological Education Model in Early Childhood Islamic Religious Learning Based on Creative Imagination. The integration develops three themes: **the potential for learning with nature, neuroscience-based Islamic education, and the development of the creative imagination of students, especially in early childhood**. *First* is the potential for learning about nature; as previously stated, the current environmental crisis is worsening. The potential for crises will continue to increase if it is not handled immediately, especially in Bangka Belitung, which has many mining areas. Yusmaliana et al. (2022b) cited Yuliana (2016), revealed that mining, such as tin mining, will harm the environment if it is not appropriately managed. It may even encroach on protected forests, conservation areas, and watersheds (DAS), leading to

Fig. 2 Teachers help the students in playing and get the meaning of learning

consequences such as flooding, drought, erosion, ecosystem damage, the spread of malaria, and landslides.

Second, neuroscience-based Islamic education in looking at nature reinforces values specifically oriented to preventing damage to nature. Suppose in the brain area; six parts regulate character. In that case, the character education of loving the environment is the proper Islamic education to become *the caliph fil ard*. Neuroscience is a basic science that can be used to analyze various contemporary Islamic education issues today (Suyadi et al. 2021). *Third*, creative imagination in early childhood can be developed to maximize the potential of the brain contained in students—especially students still in the very rapid brain development stage, namely at an early age. The rapid development of the brain in the womb, from about 100 billion nerve cells to thousands of trillions, will continue to grow in the first years of life (Kesuma and Istiqomah 2019). Therefore, maximizing this potential must be done to the maximum extent possible so that the behavior of loving the environment can be instilled as it is in the nature of a human being to become a *caliph fil ard*.

"*He arranges (yudabbiru) affairs from heaven to earth, then (affairs) go up to Him in one day, whose level (length) is a thousand years according to your calculations*" (Q.S. Al-Sajdah: 05). The cosmos stands as a testament to the magnificence of Allah SWT, and the aforementioned passage highlights that Allah created humanity to fulfill the role of caliphs on Earth, entrusted with the responsibility of regulating and managing it to the best of their abilities (Zakaria and Yusmaliana 2023).

The design of the learning model that incorporates Islamic education, environmental education, and creative imagination in early childhood can be illustrated in Fig. 3 below.

The findings of this study have important implications for ecological education in Islamic schools, emphasizing the value of integrating creative imagination into the curriculum. The positive outcomes observed in terms of enhanced understanding of environmental concepts, increased engagement in sustainable practices, and heightened environmental awareness highlight the potential of this approach to foster ecological literacy among students. These findings can inform curriculum development and educational practices not only in Islamic schools but also in other educational contexts, encouraging the incorporation of creative and imaginative strategies to enhance environmental education.

Based on the results, several practical recommendations can be made. Firstly, ongoing professional development opportunities should be provided to teachers to strengthen their pedagogical skills in integrating creative imagination in ecological education. This can include workshops, training sessions, and collaborative lesson planning. Secondly, curriculum designers should consider the inclusion of project-based learning activities that encourage students to apply their knowledge and imagination in addressing real-world environmental challenges. Thirdly, fostering partnerships with local environmental organizations and experts can provide students with authentic experiences and role models, further enhancing their environmental engagement and awareness. Lastly, continuous evaluation and feedback mechanisms

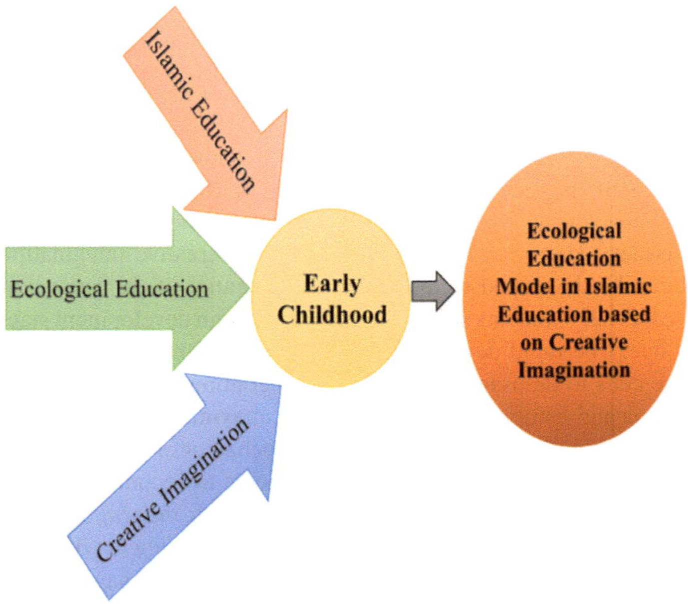

Fig. 3 Insertion of ecological education model in early childhood Islamic Religious learning based on creative imagination

should be established to monitor the effectiveness of the integration of ecological education and creative imagination, ensuring its sustained impact on students' environmental literacy and engagement.

5 Conclusion

In conclusion, this study has shed light on the experiences and perceptions of teachers and students regarding the integration of ecological education and creative imagination at Sekolah Alam Bangka Belitung (SABB). The results demonstrate the positive impact of this integration on students' understanding of environmental concepts, engagement in sustainable practices, and overall environmental awareness. Sekolah Alam Bangka Belitung (SABB) incorporates ecology education into its subjects and has a significant impact on changing the behavior of students. All informants showed a change in attitude that adhered to Islamic values and was supported by creative imagination. These findings hold implications for ecological education in Islamic schools and offer practical recommendations to enhance the integration of creative imagination in school curricula. By continuing to explore and refine these approaches, it can contribute to the development of environmentally literate and responsible individuals who are equipped to address the environmental challenges of the time.

The ecological education model in Islamic religious learning is seen as effective because the way it works can penetrate all fields of science without reducing its substance. If environmental education is integrated into Islamic education using a creative imagination approach, it will help build a strong foundation from an early age. Because the nerves in the child's brain are in a speedy growth period, Islamic values will be more firmly embedded in a person.

The creative imagination approach can shape the behavior of loving the environment in students' minds rationally, contextually, and empirically and is no longer just a textual and normative science. The behavior of loving the environment will become an awareness of the real spirit of the embodiment of piety manifested in everyday life and is no longer limited to hadiths and verses that are echoed without implementation.

References

Abd. Aziz (2014) Konservasi Alam dalam Perspektif Etika Islam; Tantangan dan Tuntutan Globalisasi. Akademika 19(2):304–321. https://doi.org/10.36835/assyariah.v5i2.116

Asroni A (2021) Pendidikan agama Islam berperspektif Ekologi. Angew Chem Int Edn 6(11):951–952, 18(85):2013–2015

Baro'ah S, Qonita SM (2020) Penanaman CiLi (Cinta Lingkungan) Pada Siswa Melalui Program Lingkungan Sekolah Tanpa Sampah Plastik. J Pancar: Pendidikan Anak Cerdas Dan Pintar 4(1):11–16

Barrett FD (2001) Creativity techniques: yesterday, today, and tomorrow. S.A.M. Adv Manag J 25–35

Bashyroh U, Mahmud A (2021) Keseimbangan Ekologis dalam Tafsir al Misbah (Studi Analitik Peran Manusia terhadap Lingkungan). SUHUF 33(2):218–231

Basri H (2013) Landasan Pendidikan. CV Pustaka Setia

Creswell JW (2012) Educational research: planing, conducting, and evaluating qualitative and quantitative, 4th edn. Pearson Edcation Inc.

Delmotte VM, Zhai P, Pirani A, Connors SL, Pean C (2021) Technical summary. contribution of working Group I to the sixth assessment report of the intergovernmental panel on climate change. In: Climate change 2021: the physical science basis. IPCC. www.ipcc.ch

Faaizun AN (2014) Model Pembelajaran Rasulullah SAW dalam Perspektif Psikologi. Pendidikan Agama Islam XI(1):19–36

Fan K, Xiao P (2015) The effects of learning styles and meaningful learning on the learning achievement of gamification health education curriculum. Eur J Math Sci Technol Educ 11(59):1211–1229. https://doi.org/10.12973/eurasia.2015.1413a

Gunarti W (2013) Sains dan Kreativitas: Pandangan Terhadap Pemikiran John Horgan dalam "The End of Science." J Desain 1(01):61–68

Heath G (2008) Exploring the imagination to establish frameworks for learning. Stud Philos Educ 27(2):115–123. https://doi.org/10.1007/s11217-007-9094-7

Indra W, Fitria Y (2021) Pengembangan media games IPA Edukatif Berbantuan Aplikasi Appsgeyser Berbasis Model PBL untuk Meningkatkan Karakter Peduli Lingkungan Siswa Sekolah Dasar. Jurnal Edukasi Matematika Dan Sains 9(1):59–66. https://doi.org/10.25273/jems.v9i1.8654

Kesuma U, Istiqomah K (2019) Perkembangan Fisik dan Karakteristiknya serta Perkembangan Otak Anak Usia Pendidikan Dasar. Madaniyah 9(2):217–236

Kholis N, Aulassyahied Q (2022) Understanding ecological Hadiths with interconnective approaches in responding to ecological problems. IJISH (Int J Islamic Stud Hum) 5(1):25–41

Lima J (2018) Learning science through art, imagination, and reflection. Creat Acad Mag 11A:33–36

Moechiji SDN (2022) Interview result with the vice principal of SABB

Muhaimin (2015) Membangun Kecerdasan Ekologis (Model Pendidikan untuk Meningkatkan Kompetensi Ekologis). Alfabeta

Nizar FA (2021) Pengaruh Manajemen Pendidikan Karakter dan Budaya Religius terhadap Kesalehan Sosial Siswa di MI Negeri Kabupaten Grobogan. Wahana Akademika: J Studi Dan Sosial 8(1):33–55

Nurhayati A, Ummah ZI, Shobron S (2018) Kerusakan Lingkungan dalam Al-Qur'an. Suhuf 30(2):194–220. http://journals.ums.ac.id/index.php/suhuf/article/view/7643

Pasiak T (2010) Pendidikan Karakter Sebagai Pendidikan Otak. J Akrab: Aksara Agar Berdaya 1(3):7–15

Pasiak T (2007) Brain management for self improvement. PT. Mizan Pustaka

Pelaprat E, Cole M (2011) "Minding the Gap": imagination, creativity and human cognition. Integr Psychol Behav Sci 45(4):397–418. https://doi.org/10.1007/s12124-011-9176-5

Pr ABP (2022) Kesalehan Sosial yang Ekologis. Tribun Jateng 1–5. https://jateng.tribunnews.com/2022/04/02/opini-aloys-budi-purnomo-pr-kesalehan-sosial-yang-ekologis?page=all

Dijksterhuis, A. (2004). Think different: The merits of unconscious thought in preference development and decision making. *Journal of Personality and Social Psychology, 87*(5), 586–598. https://doi.org/10.1037/0022-3514.87.5.586

Sari WP, Aprilliandari DI (2021) The development of PBL integrated bioenvironmentalscience app (BES app) for environmental learning. Bioeduscience 5(3):264–271. https://doi.org/10.22236/j.bes/537799

Schousboe I, Winther-Lindqvist D (2013) Children's play and development: cultural-historical perspectives. Springer Dordrecht, Heidelberg, New York, London. https://doi.org/10.1007/978-94-007-6579-5

Shalihin RR, Yusmaliana D, Suyadi, Faturahman D (2021) Rational brain transmutation into intuitive brain based on neuroscience in Islamic education. In: BICED 2020: proceedings of the 2nd EAI Bukittinggi international conference on education, BICED 2020, Bukititinggi, West Sumatera, Indonesia. European Alliance for Innovation

Smith JA, Flowers P, Larkin M (2009) Interpretative phenomenological analysis: theory, method and research, no May 2014. SAGE Publications Ltd. http://books.google.com/books?id=WZ2Dqb42exQC&pgis=1

Sombrio GDS, Ulbricht VR, Haeming WK (2014) Games and gamification : a proposal for a creative learning process in education. J Educ Hum Dev 3(4):117–129. https://doi.org/10.15640/jehd.v3n4a12

Sousa DA (2012) Bagaimana Otak Belajar: Edisi Keempat. Tim Indeks

Sulistyo A (2018) Konsep Pendidikan Lingkungan Hidup Dalam Pandangan Islam. Cahaya Pendidikan 4(1):45–58. https://doi.org/10.33373/chypend.v4i1.1281

Suyadi, Nuryana Z, Asmorojati AW (2021) The insertion of anti-corruption education into Islamic education learning based on neuroscience. Int J Eval Res Educ 10(4):1417–1425. https://doi.org/10.11591/IJERE.V10I4.21881

Suyadi (2018) Kisah (Storytelling) pada Pembelajaran Anak Usia Dini dalam Kajian Neurosains Pendidikan Islam. Jurnal Ilmiah Islam Futura 18(1):52–74. https://doi.org/10.22373/jiif.v18i1.3130

Syahril N (2022) Interview result with the Head Master of School at SABB

Syamsudin M (2017) Krisis Ekologi global Dalam Perspektif Islam. Jurnal Sosiologi Reflektif 11(2):83. https://doi.org/10.14421/jsr.v11i2.1353

Thambu N, Prayitno HJ, Zakaria GAN (2021) Incorporating active learning into moral education to develop multiple intelligences: a qualitative approach. Indones J Learn Adv Educ (IJOLAE) 3(1):17–29. https://doi.org/10.23917/ijolae.v3i1.10064

Tyan B (2005) Melatih Anak Berpikir seperti Jenius: Menemukan dan Mengembangkan Bakat yang Ada pada Setiap Anak. PT. Gramedia Pustaka Utama

Wantini, Suyatno, Dinni SM, Yusmaliana D (2022) Learning model during post covid-19 pandemic to improve early childhood language development. Jurnal Pendidikan Anak Usia Dini 6(2):6187–6197. https://doi.org/10.31004/obsesi.v6i6.3115

Watsiqotul DKK (2018) Peran Manusia Sebagai Khalifah Allah di Muka Bumi Perspektif Ekologis dalam Ajaran Islam A. Pendahuluan umat didunia, khususnya masyarakat Indonesia. Saat ini, Islam. Jurnal Penelitian 12(2):355–378. https://journal.iainkudus.ac.id/index.php/jurnalPenelitian/article/view/3523/pdf

Widiawati N (2019) A study of the epistemological thought of Al-Farabi. Al-Afkar J Isalmic Stud 2(1):48–63. https://doi.org/10.5281/zenodo.2546854

Yusmaliana D, Suyadi (2019a) Imajinasi Kreatif Berbasis Neurosains Pada Pembelajaran Pendidikan Islam. Edukasia: Jurnal Penelitian Pendidikan Islam

Yusmaliana D, Suyadi (2019b) Pengembangan Imajinasi Kreatif Berbasis Neurosains dalam Pembelajaran Keagamaan Islam. Edukasia: Jurnal Penelitian Pendidikan Islam 14(2):267–296. https://doi.org/10.21043/edukasia.vi4i2.4213.

Yusmaliana D, Suyadi S, Widodo H, Suryadin A (2020) Creative imagination base on neuroscience: a development and validation of teacher's module in covid-19 affected schools. Univ J Educ Res 8(11B):5849–5858. https://doi.org/10.13189/ujer.2020.082218

Yusmaliana D, Sabri F, Fitriana F (2022a) Pendampingan Kelompok Tanggap Bencana dengan Pendekatan Fikih Kebencanaan Sebagai Mitigasi Bencana Alam di Desa Batu Beriga , Kabupaten Bangka Tengah. Jurnal Pengabdian Pada Masyarakat 7(1):87–96. https://doi.org/10.30653/002.202271.30

Yusmaliana D, Suyadi, Widodo H, Zakaria GAN (2022b) Planning framework dan outline model Imajinasi Kreatif Berbasis Neurosains (IKBN) pada Pembelajaran Keagamaan Islam Tingkat Sekolah Dasar. Edukasi: Jurnal Penelitian Pendidikan Agama Dan Keagamaan 20(2):233–251. https://doi.org/10.32729/edukasi.v20i2.764

Yusmaliana D, Widodo H (2019) Reconstruction of Islamic education curriculum in the disruption era. International Journal of Islamic Studies and Humanities 2(1):50–57

Zakaria GAN, Yusmaliana D (2023) Management of education boarding schools 1. EDUSOSHUM J Islamic Educ Soc Hum 3(1):50–58

Sustainable Peacebuilding Through a Dignity Lens: A Case Study of Caste-Based Discrimination in Nepal

Ravi Shankar Rajaratnam and Mokbul Morshed Ahmad

Abstract The 2030 Agenda of the United Nations for Sustainable Development is undoubtedly a rights-based programme linked to conflict prevention and resolution. Dignity is strongly associated with human rights and a human rights-based approach to development. Further, dignity becomes a core issue in the peacebuilding discourse. Social cohesion is vital to peacebuilding by addressing social tensions, power imbalances, and structural inequalities. The caste system determines many people's identity and social status in Nepal, where caste-based discrimination is still prevalent. This paper aims to surface the issue of caste-based discrimination as a barrier to promoting social cohesion that would hinder peacebuilding at a societal level. This study looked at how adolescent girls perceive dignity. Hence, in-depth interviews were conducted with sixty Dalit adolescent girls in Nepal. The data and information were collected and analysed using grounded theory methodology. The findings confirm that caste-based discrimination is a violation of dignity. Adolescent girls tried to avoid upper caste people and expressed anger against them, which hampers social cohesion. Issues related to caste-based inequality, as a source of conflict, should be prioritised in the development agenda to promote social cohesion. When dignity is violated, it can cause conflict, disrupt peace, and hamper sustainable development. Hence, peace at a societal level cannot be guaranteed unless dignity is considered a core issue in peacebuilding initiatives to achieve the desired results. The authors believe that the preliminary findings will assist development practitioners in focusing their peacebuilding initiatives through a dignity lens to maximise the expected results.

Keywords Human dignity · Caste-based discrimination · Dalit · Peacebuilding

R. S. Rajaratnam (✉) · M. M. Ahmad
Development Planning Management and Innovation (DPMI) in the School of Environment, Resources and Development, Asian Institute of Technology, Pathum Thani, Thailand
e-mail: rsrajaratnam@outlook.com

A. Sharifi et al. (eds.), *Bridging Peace and Sustainability Amidst Global Transformations*, World Sustainability Series,
https://doi.org/10.1007/978-981-99-7572-3_5

1 Introduction

The sustainability-peace nexus is crucial to advancing humanity globally and is a vital component of the global agenda for sustainable development (Amadei 2021). The United Nations 2030 agenda clearly articulates, "There can be no sustainable development without peace and no peace without sustainable development" (United Nations 2015, p. 2). Hence, with SDG 16, peace is completely incorporated into the sustainable development agenda and is a cross-cutting issue regarding gender equality, governance, health, inequalities, security, assistance for vulnerable nations, and sustainable cities (Amadei 2021). However, the Sustainable Development Goals are unlikely to be achieved since conflict is a severe development obstacle (Yarnall et al. 2021).

The primary goal of the 2030 Agenda is to "leave no one behind," a promise anchored in the human rights principle of equality and non-discrimination, which is also at the heart of conflict prevention and peacebuilding (UNHCHR 2020). Inequalities are a significant source of conflict; therefore, reducing them and increasing social capital are two ways to improve horizontal and vertical cohesion, which is now recognised as essential to promoting peace and sustainable development (Löhr et al. 2022). Thus, conflict prevention and resolution are inextricably linked to sustainable development (Löhr et al. 2022).

The United Nations' 2030 Agenda for Sustainable Development is undoubtedly a rights-based programme (Choondassery 2017). Although many NGOs practise rights-based approaches to development, there is no universally agreed definition, though there is some consensus on some elements of it (O'Leary 2017; Nelson and Dorsey 2018). A human rights-based approach emphasises inclusive and meaningful participation in which different groups and communities, including ethnic and minority groups, are represented and encouraged to actively participate in processes and decision-making without discrimination (UNHCHR 2020). The United Nations acknowledges that development, peace, security, and human rights are inextricably intertwined and mutually reinforcing (Löhr et al. 2022). Hence, peacebuilding requires a rights-based development approach.

According to Choondassery (2017), the argument for the foundation of the "human rights-based approach comes from our moral nature that values the inherent dignity of the person and the moral demand of promoting and protecting it" (p. 20). This confirms how dignity is strongly associated with human rights and a human rights-based approach to development. Song (2015) interprets that the major organisations implementing programmes in line with the rights-based approach complement a human dignity approach. Similarly, scholars have claimed that many social problems, particularly those influencing health, such as violent conflict, human rights violations, and social unfairness, can be addressed through a dialectic of dignity (Perry 2013). Thus, dignity becomes a core issue in the rights-based approach to development and the peacebuilding discourse.

Human rights have long been seen as powerful moral principles that can determine the quality of peace by dealing with social tensions, power inequalities, and structural inequality (Georgi 2022). In societies dominated by systemic inequalities, incompatible values lead to conflicts of interest (Matijević and Erić 2015). Hence, social cohesion is essential in addressing social tensions, power, and structural inequalities (Georgi 2022). Over the last two decades, governments and international donor organisations have increasingly recognised the importance of social cohesion in peacebuilding, recognising the importance of social fragmentation and establishing trust between social groups and the state as critical for ensuring peacebuilding (Löhr et al. 2022). This paper aims to surface the issue of caste-based discrimination as a barrier to promoting social cohesion that would hinder peacebuilding at a societal level.

Inherited caste identity determines one-fifth of the world's population's life opportunities, but it is not prioritised in global policy debates (Mosse 2018). For many people in Nepal, the caste system determines "their identity, social status, and life changes" (Bennett et al. 2008, p. 1). Dalits in Nepal are subjected to discrimination and exclusion due to a hierarchical caste system where Dalits are the lowest in the system (Doss et al. 2022). The socio-economic conditions of the Dalits are lower as compared to other ethnic and caste groups in Nepal (Bhattachan et al. 2009). Although Nepal is trying to move away from the caste system, caste-based discrimination against Dalits is still prevalent. Human dignity is linked to social inclusion and freedom and the "absence of discrimination, shame, and humiliation" (Hojman and Miranda 2018, p. 2). Accordingly, this paper focuses on determining how Dalit adolescent girls view caste-based discrimination and their dignity, which are affecting social cohesion and, eventually, peacebuilding at the community level.

2 Literature Review

The word dignity derives from the Latin '*Dignitas*', meaning honour and respect (Staffen and Arshakyan 2017). It became a focal point of discussion in philosophy when the German philosopher Immanuel Kant defined dignity as intrinsic to human beings (Loughlin 2016). Later, it became a central concept in the development and legal sectors when the United Nations recognised dignity in the Universal Declaration of Human Rights (UDHR) preamble.

Comprehending the multifaceted concept of dignity remains a challenge, with scholars, politicians, and philosophers all defining it differently. From the Kantian perspective, dignity is intrinsic and inalienable. However, human dignity is culturally relative, dependent on time and place, and dependent on the circumstances of a given scenario (Kleindienst 2017). In some situations, dignity is considered a quality of interaction between individuals and how they are treated; in other situations, it is a psychological or cognitive outcome that allows people to achieve a 'sense of dignity' (Keisu 2017). Thus, dignity can be said to be composed of at least two dimensions; one is acquired at birth by virtue of being born as a human being, and the other is acquired as the result of certain behaviours and the social environment (Staffen and Arshakyan 2017). Similarly, Jacobson et al. (2009) suggest that human dignity has two distinguishable features: intrinsic and social dignity, where the latter results from interactions between individuals, collectives, and societies. Hence, according to Jacobson et al. (2009), social dignity may be promoted or violated through encounters between and among actors.

Caste is a social classification based on birth within a group (Dhanda 2020). According to Patra and Velassery (2013), "casteism is human alienation" (51). Dalits in Nepal are subjected to discrimination and exclusion due to a hierarchical caste system where Dalits are the lowest in the system (Doss et al. 2022). Dalits are still among Nepal's most economically disadvantaged, politically marginalised, and socio-culturally oppressed groups (Pariyar and Lovett 2016). Participants in a study conducted by Pariyar and Lovett (2016) expressed high anxiety about the term "Dalit" as it has long-term implications for their identity. Thus, casteism should be considered a human rights issue that is discussed from a global perspective and graded on the same scale as other human rights issues (Dhanda 2020).

Researchers on peace and conflict frequently use Johan Galtung's work on violence and peace as a starting point. He believed peace was not merely the absence of war but also structural and more invisible forms of violence (Sharifi et al. 2021). Hence, a comprehensive definition of peace includes freedom from less obvious kinds of violence, such as social discrimination, political repression, and other structural injustices that impede the growth of individual agency and opportunity (Fisher et al. 2021). Therefore, for establishing and maintaining peace, academics emphasise the importance of minimising power disparities (Sharifi et al. 2021).

A peaceful, democratic, and wealthy society is said to be strongly influenced by social cohesion (Löhr et al. 2022). Lack of social cohesion elevates inequality and discrimination in a given context. Horizontal cohesion, such as relationships among individuals of the same group and relationships among individuals across groups, and vertical cohesion, such as relationships between individuals, groups, and the state, are crucial to ensuring sustainable development and peace (Löhr et al. 2022). Muller and Neuhäuser (2011) argue that poor people see themselves as 'second-class citizens', which undermines their self-respect and, accordingly, their dignity. Similarly, Dalits in Nepal too consider themselves second-class citizens as they are rated as outcasts outside of the caste system, and therefore they became untouchables (Patra and Velassery 2013).

The new constitution of Nepal highlights "the right against untouchability and caste-based discrimination and 'the rights of Dalits' as fundamental rights" (Committee on the Elimination of Racial Discrimination 2017, p. 4). However, caste-based discrimination still prevails in Nepal even after the Nepalese government has introduced different acts and actions. According to Galtung, if injustice and repression are embedded in the foundations of a society, it will eventually evolve into structural violence (Matijević and Erić 2015). Hence, Galtung believes that peace-building, rather than traditional peacekeeping and peacemaking, is the only method to address these structural roots of violence and resolve the conflict (Matijević and Erić 2015).

3 Methods

Our study examines young people's understanding of dignity as revealed by how it is violated among adolescent Dalit girls in Nepal. This article is mostly based on qualitative research. The data was collected and analysed using grounded theory methodology. Sixty adolescent girls were questioned using a semi-structured interview guide. This study focused on Dalit teenagers aged 15–19 to collect rich qualitative data. The data collection started in February 2020, after getting approval from AIT's Research Ethics Review Committee, and ended in February 2021.

This study was undertaken in the Kabilvastu district, where a local NGO implemented an adolescent empowerment programme in line with the principles of a rights-based approach to development. Adolescent girls who met our criteria were purposefully selected. For the study, a female Dalit research assistant was hired who had local and English communication skills, was experienced with data collection, and did not have affiliations with any NGOs in Nepal. She assisted the adolescent girls in recalling various elements of their lives and questioned them on how such experiences honoured or violated their human dignity. The interviews took place without the presence of NGO staff. In accordance with research ethics, all participants provided prior written consent. The qualitative analysis was supported by the use of MAXQDA software.

4 Results

Most of the girls reported being humiliated in common places, mainly when they went to collect water at the collection points. Some of the responses from the girls follow a similar pattern. Adolescent girl 1 (18 years old) said that:

> Whenever we went near the handpump to fill water, the upper-caste woman used to ask us to stay away from this place. They say they want to fill the pots first, and then we can take the water. Once she said, "Get out of this place, I had to fill the water tank first". Mistakenly, if we touched their pots, they started rubbing them with detergent powder and scolding us too. (Adolescent Girl 1, 18 years old)

The Adolescent Girl 12 (15 years old) reiterated that:

> When I went to the tap, they (upper-caste women) said to stay away. Once, when I kept my pot and started filling it first, they pounced on it and started filling theirs. I felt terrible and could not eat anything, so I went to bed without eating. (Adolescent girl 12, 15 years old)

Almost all girls expressed similar experiences about fetching water at the handpump. Adolescent girls reported feeling humiliated and avoiding going to fetch water when upper-class women were present at the water collection places. Adolescents reported issues affecting their day-to-day lives, mainly with upper-caste older women, and half of the adolescents interviewed reported humiliation at the water collection points. In addition, adolescents also reported being ill-treated by their friends. Adolescent girl 25 (16 years old), expressed her feelings when her friend attempted to discriminate against her because of her caste.

> When we were in Pyuthan (a neighbouring district), I was going to my friend's house one day. We went there with five other friends. Then we became thirsty after hiking. One of my friends asked for water from an unknown house. They gave us some, but one of my friends told me, "I will drink first, then you," and I asked why. She said, "You are of a lower caste, so we should not eat or drink anything if you touch it.". At that time, I felt bad for myself for being born into a lower-caste family. (Adolescent girl 25, 16 years old)

In this manner, all adolescents reported facing challenging situations with upper-caste people, mainly older women and their upper-caste friends. They mostly encountered such situations when visiting upper-caste families and friends, which resulted in negative feelings. They were hurt by incidents in the relationship between the upper and lower castes and others' recognition of their status. Adolescent girl 8 (16 years old) shared her feelings when an upper-caste shop owner maltreated her.

> Their (higher caste people's) words have the power to break someone's heart. When I was young and went to buy chocolate in a small boutique nearby, I mistakenly touched the chocolate box. The shop owner then slapped me so hard. I could not understand why she slapped me. Later, I asked my mother, and only I knew it was because of our caste. (Adolescent girl 8, 16 years old)

The caste-based discrimination contributed to mixed feelings among adolescent girls. First, the frustration of being born in the Dalit community was highlighted as Adolescent girl 7 (15 years old, school dropout) said, "I feel bad; if I were born in another caste, they would have treated me equally, but now because of my caste, they treated me differently." The difference in status pushes adolescent girls to articulate equality. Her desire to be born into an upper-caste family grows, and she believes her social standing will improve. On the other hand, since status becomes an attribute of dignity, she indirectly touched upon her dignity. Secondly, adolescent girls expressed their anger towards upper-caste women and their friends. Adolescent girl 11 (17 years old) expressed her multiple feelings due to her experience with upper-caste women.

> When she (upper caste woman) said, You are Dalit, and we are upper class, do not come inside our house. Then I felt terrible and became enraged. I wish I could have beaten her. Then I started to unfollow my friendship with her daughters, and now their daughters are not my friends. I felt angry when they started scolding me for things I was helpless or not responsible for. I once went 4-5 days without eating due to stress and a headache. (Adolescent girl 11, 17 years old).

Money and resultant power, traditions, rules and regulations, beliefs about curses, and the caste system itself contribute to the superior behaviour reported by adolescent girls, as shown in Table 1.

Table 1 Reasons for the superior behaviour of upper-caste women

Reason	Sample quote
Money and power	"Their supremacy over everything, including money, power, and strength. So, they always think that Dalits should always be below them." (Adolescent girl 42, 15 years old)
Tradition, rules, and regulations	"I think their tradition or rules and regulations contribute to their thinking of themselves as superior to others." (Adolescent girl 45, 19 years old)
Beliefs of the curse	"They teach their children that by touching a Dalit they will get a curse from their forefathers and ancestors." (Adolescent girl 57, 19 years old)
Caste system itself	"Their supremacy over other castes, according to Dharma is like that of the Bharhmin, followed by the Khsyatriya, Vaisya, and Sudra. As a result, they are always convinced that they are superior to others." (Adolescent girl 49, 18 years old)

5 Discussion

The negative feelings of adolescent girls helped unpack the concept of dignity in their context. These feelings did not hurt them physically but hurt the adolescent girls mentally, especially in their subconscious minds. Kant believes that "human beings have an inner transcendental kernel", which is the basis for their dignity (Kant, as cited in Loughlin 2016, p. 328). Hence, these feelings could be attributed to Kant's inner transcendental kernel. In other words, it could be argued that the feelings are the result of treating them with indignity. Human dignity, intrinsic dignity, and social dignity, which result from interactions between and among individuals, collectives, and societies, are considered distinct phenomena by scholars (Jacobson et al. 2009). Adolescent girls also experienced a lack of social dignity due to their interactions with the so-called upper caste people in their community. Hence, caste-based discrimination becomes one of the issues contributing to the violation of the dignity of adolescent girls.

Although most of the girls mentioned that the elderly women in their society mainly practise caste-based discrimination, as they are referred to as 'aunties and grandmothers', they also experienced it through their friends. Adolescent girls also shared incidents in which their friends humiliated them, resulting in anger and withdrawal from such incidents. Almost two-thirds of the girls said they started avoiding situations, such as not going to fetch water when upper-caste women are present at the tube-water collection place, and girls also said they stopped having relationships with and visiting their upper-caste friends. These practices mainly avoid potential conflicts between the Dalits and the upper caste.

The main reason for breaking the relationship with the upper-caste people is to avoid situations that could trigger violence or humiliate the adolescent girls and preserve their dignity; as Kant says, the inner transcendental kernel is the basis of their dignity. Adolescent girls also expressed the reason why upper-caste people treat them differently. The common factor that surfaced during the interviews is that upper-caste people's superior mentality largely influences their behaviour. Adolescent girl 30 (16 years old), expressed this clearly.

> The main factor causing them to behave like that is probably their traditional way of thinking, what they have learned from their parents, and what they teach their children. Their psychology means that they think they are superior to others. (Adolescent girl 30, 16 years old)

Some people believed they were superior to others because of their perceived caste status. Hence, the caste system and the resultant community practises demarcating the 'upper' and the 'lower', which influences the girls' perception that they are inferior in their community. Adolescent girls' understanding of superior behaviour is governed by status. Adolescent girls stated different reasons for the superior behaviour of upper-caste people (Table 1). According to adolescent girls 42 and 49, upper-caste people's superiority made them believe they were lower than them. Mosse (2018) articulates that caste as a social system was established when "the superiority of Brahman purity over Kshatriya power established the ideological separation of status and power" (425). Adolescent 45 reflected this by saying that the traditional rules and practises contributed to the feeling of superiority. As a result, because it is a privilege for Brahmans, the upper caste people are given advantages; they are entitled to honour, respect, and precedence in all social matters (Patra and Velassery 2013).

The superior thinking of upper caste people discriminates against Dalit adolescent girls. This has created a clear demarcation between the lower and upper castes, as expressed by the adolescent girls. The quotes by them, such as "I am afraid of them. Because they are from a higher caste." (Adolescent girl 17, 15 years old) and "I see myself lower than the other caste when they (upper caste) treat me differently." (Adolescent girl 27, 15 years old) confirms a clear demarcation in status due to the caste system. Caste has an ideological dimension associated with ritual ranking and ideas of purity and impurity (Mosse 2018). Hence, the upper caste people think the lower caste people are impure. Further, the Dalits are rated as outcasts outside of the caste system, and therefore they became untouchables (Patra and Velassery 2013). Such demarcation, dominated by systemic inequalities, impedes community peace.

Dalits in Nepal are subjected to discrimination and exclusion due to a hierarchical caste system where Dalits are the lowest in the system (Doss et al. 2022). Similarly, Dalit adolescent girls also reported being lower than others. The feeling of thinking 'low' is associated with dignity. Adolescent girl 52 (15 years old), said, "Obviously, we are Dalits, and we have a lower position in the community." Adolescent girl 46 (19 years old), assured that they are always lower than others by saying, "We were, we are, and we will always be lower than them no matter what we do." Thoughts of "being lower' are deeply inculcated in adolescent girls' minds, which makes it difficult for them to think that their status can be improved in the community and that they can also lead a life with dignity. According to adolescent girls, upper-caste people's superior thinking impacts interpersonal interactions. According to Chadwick (2012), personal values, behaviours, and attitudes can largely influence interpersonal skills, and dignity also plays a vital role in interpersonal interaction.

Hence, adolescent girls tend to discontinue their relationships, even with friends from the upper caste. Such problem-avoidance behaviour may not contribute to long-term peace and development.

Adolescent girls try to avoid circumstances where they could feel unequal, as they end up with negative feelings such as humiliation and distress. The rejected claim to equality by Dalits is "devastating, hurtful, even traumatic", and such "dignity humiliation" is a source of distress (Mosse 2018, p. 433). As a result, feelings of distress and humiliation result from a dignity violation. The interpersonal relationships, even with their friends, confirm caste-based discrimination, which contributes to multiple negative feelings confirming their lower social status. Hence, adolescent girls reported avoiding such circumstances as described, as shown in Table 2.

Adolescent girls try to avoid such circumstances, mainly to avoid being hurt and humiliated. The underlying reason for this withdrawal mechanism is twofold, one is to avoid being discriminated against due to caste, and the other is their expectation of being treated equally as human beings. In other words, they wanted to be in a zone where they felt dignified, as scholars interpret that even if a person lacks his or her "realised dignity," they will retain their "inherent dignity," which is also called "initial dignity" as it is inherent to the human being (Kleindienst 2017). Because of human beings' nature of free will, they have basic dignity, and according to Kant, human beings can realise their dignity by themselves (Giesinger 2012). In this manner, adolescent girls attempt to realise their dignity while avoiding circumstances that violate it. Adolescents isolate themselves from mainstream society in order to protect their dignity, in addition to being excluded by the upper caste. As per UN terminology, "fundamentally, caste determines social exclusion" (Mosse 2018). Hence, it was evident that adolescent girls' problem-avoidance behaviours affect social cohesion, which is crucial for peacebuilding at the community level.

Regardless of age, sex, socio-economic level, physical condition, ethnic origin, political ideals, or religion, all human beings have equal and intrinsic worth and

Table 2 Avoidance of circumstances that cause humiliation

Reason	Sample quote
Visiting friends' house	"When I visited my friend's house (upper caste), she openly said not to sit here. I did not feel good, and then I decided that I would never go to their house again." (Adolescent girl 4, 19 years old)
Visiting upper-caste people	"I know which house or which people talk about caste discrimination, but I never go to their house." (Adolescent girl 15, 17 years old)
Fetching water	"One day when I went to my neighbour's tap for water, she (of the upper caste) asked me to stay away, so I stayed away, and because of my caste, I felt humiliated." I felt sad and tense. "I never go there." (Adolescent girl 9, 17 years old)
Attending social events	"At a party or any programme held in society, 2–4 groups of people sit together and have fun, but we have to sit alone. I feel sad, and that's why we do not go to anyone's house for a party." (Adolescent girl 10, 17 years old)

should be treated with the utmost respect and care (Andorno 2014). However, upper-caste people always think they are superior due to the caste system. Due to the "idea that the upper caste people are superior due to their race, look, and colour… most of our upper caste people's approach and attitude towards the Dalits are based on a contemptuous feeling" (Patra and Velassery 2013, p. 50). As a result of their perceived power as a result of the caste system, their behaviour towards Dalits is humiliating. The salient feature of "caste and all its social judgements is a modern form of power over Dalits" (Mosse 2018, p. 433). Due to such power, as adolescent girls reported, Dalits feel humiliated and distressed. According to Patra and Velassery (2013), Dalits were part of the system because their services were critical to the community, but they were treated inhumanely. Jacobson (2009) confirms that the violation of dignity is associated with a social order of inequality. Hence, from the point of view of dignity, caste-based discrimination could be considered a violation of social dignity.

The findings confirm that caste-based discrimination, as a violation of dignity, hampers social cohesion. As adolescent girls explained, they tried to avoid the so-called upper caste people and expressed anger against them. Peacebuilding should begin with oneself and then spread to the family, community, and, ultimately, the societal or national level. However, according to our findings, hatred is being inculcated among young adolescent girls from the Dalit community. Similarly, upper-caste women and girls instilled such hatred in themselves. Thus, caste-based discrimination impacts social cohesion and acts as a ticking time bomb that could detonate at any time, causing conflict between different groups. In other words, violation of dignity as a root cause could trigger conflict and therefore be an obstacle to sustainable development. Between inequality and unsustainability, there is likely to be a vicious cycle whereby increasing one will increase the other, which will then increase the former, and so on (Neumayer 2011). A significant reduction in inequality is required to maintain political, economic, and environmental sustainability (Stewart 2014). Further, unresolved conflicts can obstruct the sustainability goal's advancement (Fisher et al. 2021). Therefore, inequality as a root cause needs to be addressed, which could promote peace and eventually contribute to sustainability.

Many organisations have long aimed to create a more just, secure, and sustainable world where everyone can live in peace and dignity (Amadei 2021). However, this study confirms that even empowerment programmes implemented in line with rights-based programming principles failed to promote dignity. The second-generation conflict theories attempted to incorporate the core causes of conflict sought in society's social, psychological, and structural structures (Matijević and Erić 2015). Further, the objectives of contributions to peace include reducing conflict's root causes, fostering social cohesiveness, and altering relationships between people (Joireman and Haddad 2023). Accordingly, the findings confirm that caste-based discrimination is a significant cause of violations of human dignity, hampers social cohesion, triggers conflict, and affects sustainable development. Hence, peace at a societal level cannot be guaranteed unless dignity is considered a core issue.

6 Conclusions

As per the in-depth interviews conducted with Dalit adolescent girls in Nepal, this research confirmed that caste-based discrimination violates human dignity and should be addressed. Since caste-based discrimination is rooted in culture and tradition, it needs a different approach. Through peacebuilding initiatives, educating young people using dignified language might change their mindset, as formal education and awareness-creation programmes do not touch upon the root causes of prevailing issues related to caste. Violation of dignity as a root cause could trigger conflict and hamper sustainable development. Promoting social cohesion is a prerequisite for sustainable peacebuilding at the community level, which cannot be achieved as per the findings unless caste-based inequalities are prioritised in the development agenda. Therefore, in order to achieve the desired results of peacebuilding initiatives, the root cause of violations of dignity should be addressed. This study focused only on Dalit adolescent girls. Hence, further studies are recommended to explore the interconnectedness of dignity and peacebuilding. However, the authors believe this preliminary finding will assist development practitioners in focusing their peacebuilding initiatives through a dignity lens to maximise the expected results.

References

Andorno R (2014) Human dignity and human rights. In: Have HT, Gordijn B (eds) Handbook of global bioethics. Springer, Dordrecht, pp 45–57. https://doi.org/10.1007/978-94-007-2512-6_66

Amadei B (2021) A systems approach to the sustainability–peace nexus. Sustain Sci 16:1111–1124. https://doi.org/10.1007/s11625-020-00902-x

Bennett L, Dahal DR, Govindasamy P (2008) Caste, ethnic and regional identity in Nepal: further analysis of the 2006 Nepal demographic and Health survey. Macro International Inc., Calverton, Maryland, USA

Bhattachan KR, Sunar TB, Bhattachan YK (2009) Caste-based discrimination in Nepal. Indian Institute of Dalit Studies. Working Paper Series, vol III

Chadwick A (2012) A dignified approach to improving the patient experience: promoting privacy, dignity and respect through collaborative training. Nurse Educ Pract 12:187–191

Choondassery Y (2017) Rights-based approach: the hub of sustainable development. Discourse Commun Sustain Educ 8(2):17–23

Committee on the Elimination of Racial Discrimination (2017) Consideration of reports submitted by States parties under article 9 of the Convention: Nepal. United Nations

Dhanda M (2020) Philosophical foundations of anti-casteism. Proc Aristot Soc 120(1):71–96. https://doi.org/10.1093/arisoc/aoaa006

Doss CR, Meinzen-Dick R, Pereira A, Pradhan R (2022) Women's empowerment, extended families and male migration in Nepal: insights from mixed methods analysis. J Rural Stud 90:13–25

Fisher J et al (2021) Four propositions on integrated sustainability: toward a theoretical framework to understand the environment, peace, and sustainability nexus. Sustain Sci 16:1125–1145. https://doi.org/10.1007/s11625-021-00925-y

Georgi FR (2022) Peace through the lens of human rights: mapping spaces of peace in the advocacy of Colombian human rights defenders. Political Geogr 99. https://doi.org/10.1016/j.polgeo.2022.102780

Giesinger J (2012) Kant on dignity and education. Educ Theory 62(6):609–620
Hojman DA, Miranda A (2018) Agency, human dignity and subjective well-being. World Dev 101:1–15. https://doi.org/10.1016/j.worlddev.2017.07.029
Jacobson N (2009) A taxonomy of dignity: a grounded theory study. BMC Int Health Hum Rights 9(1). https://doi.org/10.1186/1472-698X-9-3
Jacobson N, Oliver V, Koch A (2009) An urban geography of dignity. Health Place 15(3):725–731. https://doi.org/10.1016/j.healthplace.2008.11.003
Joireman S, Haddad F (2023) The humanitarian–development–peace Nexus in practice: building climate and conflict sensitivity into humanitarian projects. Curr Opin Environ Sustain 62:101272. https://doi.org/10.1016/j.cosust.2023.101272
Keisu, (2017). Dignity: a prerequisite for attractive work in elderly care. Soc Health Vulnerability 8:1322455. https://doi.org/10.1080/20021518.2017.1322455
Kleindienst P (2017) Understanding the different dimensions of human dignity: analysis of the decision of the constitutional court of the Republic of Slovenia on the >>Toto Streat<< case. Law Econ Rev 8(3):117–137
Löhr et al (2022) Integrating the concept of peacebuilding in sustainability impact assessment. Environ Impact Assess Rev 95. https://doi.org/10.1016/j.eiar.2022.106803
Loughlin J (2016) Human dignity: the foundation of human rights and religious freedom. Memoria Y Civilizacion 19:313–343
Matijević MV, Erić VC (2015) Peacebuilding and the conflict resolution theories. In: Đorđević et al (eds) Twenty years of human security: theoretical foundations and practical applications, Belgrade, p 151–161
Mosse D (2018) Caste and development: contemporary perspectives on a structure of discrimination and advantage. World Dev 110:422–436
Muller J, Neuhäuser C (2011) Relative poverty. In: Kaufmann P et al (eds) Humiliation, degradation, dehumanisation: human dignity violated. Springer, New York, 159–172. https://doi.org/10.1007/978-90-481-9661-6
Nelson PJ, Dorsey E (2018) Who practices rights-based development? A progress report on work at the nexus of human rights and development. World Dev 104:97–107
Neumayer E (2011) Sustainability and inequality in human development. Human Development Research Paper 2011/04. UNDP. https://ssrn.com/abstract=2351453
O'Leary S (2017) Grassroots accountability promises in rights-based approaches to development: the role of transformative monitoring and evaluation in NGOs. Account Organis Soc 63:21–41
Pariyar B, Lovett JC (2016) Dalit identity in urban Pokhara, Nepal. Geoforum 75:134–147
Patra R, Velassery S (2013) Casteism: onto-genesis of an ideology of exclusion. Int J Hum Religion 2(2):44–54
Perry DJ (2013) Peace through a healing transformation of human dignity possibilities and dilemmas in global health and peace. Adv Nurs Sci 36(3):171–185
Sharifi A et al (2021) The sustainability–peace nexus: why is it important? Sustain Sci 16:1073–1077. https://doi.org/10.1007/s11625-021-00986-z
Song AM (2015) Human dignity: a fundamental guiding value for human rights approach to fisheries? Mar Policy 61:164–170
Staffen MR, Arshakyan M (2017) About the principle of dignity: philosophical foundations and legal aspects. https://doi.org/10.5007/2177-7055.2017v38n75p43
Stewart F (2014) Sustainability and inequality. Development 57(3–4):344–361. https://doi.org/10.1057/dev.2015.1
United Nations (2015) Transforming our world: the 2030 Agenda for sustainable development. https://doi.org/10.1891/9780826190123.ap02
UNHCHR (2020) Thematic paper: The contribution of human rights to peacebuilding and sustaining peace. 2020 Review of the UN Peacebuilding Architecture. UN Office of the High Commissioner for Human Rights

Yarnall K, Olson M, Santiago I, Zelizer C (2021) Peace engineering as a pathway to the sustainable development goals. Technol Forecast Soc Chang 168:120753. https://doi.org/10.1016/j.techfore.2021.120753

Bangsamoro Youth in Peacebuilding: Contributions, Opportunities and Challenges

Yasmin Abdurahim Tagorda and Francisco A. Magno

Abstract Numerous development initiatives have been established to support youth leadership in peacebuilding programs worldwide. In the Philippines, development assistance has promoted youth participation in the Bangsamoro Autonomous Region for Muslim Mindanao (BARMM). Acknowledging that youth participation fosters peacebuilding, this study aims to examine the engagements and contributions of Bangsamoro youth civil society organizations (CSOs) in Mindanao peacebuilding. Using an institutional approach, this study seeks to identify the enabling environment for the participation of the youth in their quest to become active agents of positive change and peacebuilding. Data are obtained from CSO youth members and other peace actors using a combination of qualitative methodologies. The results suggest that the institutionalization of venues for involvement and the availability of projects given by various funding sources foster robust youth CSO engagement in the BARMM. Respondents cited community experience and school-based leadership training as motivators for joining and establishing CSOs. The establishment of networks among the Bangsamoro youth groups promote participation. However, intergenerational disparities hinder youth participation.

Keywords Bangsamoro youth · Civil society organizations · Peacebuilding

1 Introduction

The Bangsamoro Region, situated in the southwestern portion of Mindanao, Philippines, and predominantly inhabited by Muslims, has experienced protracted lethal conflicts interspersed with sporadic endeavors at negotiations spanning five decades.

Y. A. Tagorda (✉)
Graduate Student, Department of Political Science and Development Studies, DLSU, Manila, Philippines
e-mail: yasmin.tagorda@msumain.edu.ph

F. A. Magno
Department of Political Science, and Development Studies, DLSU, Manila, Philippines

A. Sharifi et al. (eds.), *Bridging Peace and Sustainability Amidst Global Transformations*, World Sustainability Series,
https://doi.org/10.1007/978-981-99-7572-3_6

This vicious cycle of war and negotiations has led to considerable devastation and adversity-claiming numerous lives, imposing high economic costs, marginalizing livelihoods, driving food shortages, and exposing communities to serious health hazards. However, there was a glimmer of optimism when the Bangsamoro Organic Law (BOL), a legislative measure establishing the Bangsamoro Autonomous Region in Muslim Mindanao (BARMM) and providing the Bangsamoro people in the area with self-governance and significant autonomy was enacted in July 2018, temporarily halting this persistent conflict. Supporting data from Conflict Alert show a 30% decrease in conflict incidence and a 60% drop in deaths in Muslim Mindanao (International Alert 2007). Despite this reduction in violence and deaths, the lingering issue of the sustainability of peace remains.

It is critical to recognize that signing the BOL is an essential step toward 'peacemaking' and 'peacekeeping'—a critical but limited commitment to halt hostilities and promote nonviolence, as Bercovitch and Kadayifci (2002) emphasized. However, achieving long-term peace necessitates a comprehensive and all-encompassing approach known as peacebuilding. At its core, peacebuilding comprises a consistent and far-reaching commitment to post-conflict reconstruction, embracing various activities and stages to transform a conflict into a stable state of peace (Lederach 1997, p. 13). As a result, it is critical to underscore the importance of ongoing involvement and interventions to address the root causes of the conflict, foster reconciliation, and preserve the region's long-term peace sustainability.

While peacebuilding is universally understood, the question of how to construct a lasting peace has been the subject of lengthy debate. Muslim Mindanao has been subject to liberal peacebuilding for quite some time now. This post-conflict agenda captures the desire of the international community to embed liberal ideas, approaches and standards, and rules toward governance via 'consensual intervention' (Richmond and Franks 2009).

The critiques of liberal peace, however, have attributed the highly centralized and outsider-driven peace process as reasons for failure (Bird 2007; Mac Ginty and Richmond 2013; Verkoren and van Leeuwen 2013; Leeuwen et al. 2020). As such, the 'local turn' has gained traction, emphasizing the importance of local ownership in the peacebuilding process. This initiative focuses on local stakeholders and promotes context and agency in establishing peace (Mac Ginty and Richmond 2013). The current stage in Mindanao peacebuilding has brought attention and appreciation to the local actors and their capabilities. The indigenous and insider approaches to managing conflict connect and correspond to what is accepted and expected by the local stakeholders. Therefore, local peacebuilding, propelled and managed by community members and other local actors, achieves more legitimacy and sustainability (Mac Ginty 2010; Boege 2011). Civil society organizations are acknowledged to have essential roles in sustaining peace (Paffenholz 2010).

Civil society offers a critical space for enhanced participation, and "diversity and pluralism can be fostered." In addition, they help guide stakeholders' responses toward constructive involvement, raising awareness of the consequences of protracted conflict and fostering dialogue (Hampson 1996; Peck 1998; Barnes 2005; UN Security Council 2005; Rood 2005; Paffenholz 2010).

There is a reasonable amount of literature about the roles of civil society in peacebuilding. Civil society actors in many forms/types (academe, church-based organizations, think tanks, NGOs, networks, CBOs, etc.) have seen engagement in the different phases of peace processes in the protracted conflict in Muslim Mindanao. However, the neglect to investigate the roles and contributions of the Bangsamoro youth civil society and the institutional circumstances that enable or hinder their interactions result in the forfeiture of significant perspectives regarding their distinct historical and cultural milieu. Examining the involvement of Bangsamoro youth in resolving the current conflict, despite their unique challenges, could provide insight into their efficacy and inventive methodologies in promoting sustainable peace. Achieving a comprehensive understanding of peacebuilding in Mindanao necessitates recognition of its unique context and contributions, which can result in more informed and targeted measures for sustainable peace.

In recognition of the local turn to peacebuilding and the subsequent support to local civil society organizations in the past 20 years, this study explores the roles and contributions of Bangsamoro youth CSOs in Mindanao peacebuilding. It also identifies the strategies they used to resist, challenge, and subvert the constraints imposed by the current peacebuilding framework.

However, before proceeding to the peacebuilding and sustainability strategies used by Bangsamoro youth civil society organizations (BMCSOs), this paper will provide a review of the relevant literature on fundamental themes such as peacebuilding, civil society, and the local turn in peacebuilding as well as a background on the Bangsamoro conflict providing important context for the study. A section dedicated to the framework for analysis will also be included. Section 2 presents the methodology used to collect and analyze the data, followed by a presentation and discussion of the study's findings. This part will be divided into three subsections, each focusing on the nature and character of the Bangsamoro Youth organizations, their substantial contributions to peacebuilding, and the contextual conditions that promote long-term peace. Finally, this article will conclude with the findings and evaluate the potential and obstacles involved in enhancing the Bangsamoro Youth organizations' pursuit of long-term peace.

1.1 Civil Society in Peacebuilding: Definitions, Concepts and Contributions

Civil Society defined. Civil society, an essential component of the peacebuilding agenda, covers a wide range of groups and is defined by several interconnected descriptions (Belloni 2001; Anheier 2004; Rooy 2004; Paffenholz and Reychler 2007). To provide a comprehensive understanding, Steffek and Nanz (2007) define civil society as all actors having legal identities recognized by at least one country, excluding official governmental bodies. This definition broadens the scope to encompass community groups, non-governmental organizations (NGOs), labor unions, philanthropic organizations, faith-based organizations, professional associations, and foundations. Voluntary and self-organized entities within civil society play an important role in organizing public interests, molding public opinion, and contributing to resolving social challenges. They are self-governing and non-profit organizations that operate independently of the state and the market. Civil society actors amplify social problems in the public domain by freely associating and pursuing shared goals, fostering inclusivity and public involvement, and promoting long-term peace (SteffekNanz 2007; Salamon 2010).

Defining Peacebuilding. Having acquired an understanding of civil society and its contribution, it becomes imperative to explore the fundamental definition and guiding principles that underpin the concept of peacebuilding. Peacebuilding is developing constructive personal, group, and political relationships across boundaries to resolve injustice nonviolently, changing the structural conditions that generate deadly conflict, and encompassing conflict prevention, management, resolution, transformation, and post-conflict reconciliation. It entails interventions and interactions between diverse players, both internal and external, to create conditions conducive to peace, give relief and reconstruction of war-torn societies, and establish institutions that encourage a sense of security (Lederach 1997). Furthermore, peacebuilding differs from related concepts, such as "peacemaking," which refers to agreements made by warring parties to halt hostilities and frequently involve diplomatic conversations and negotiations. During persistent violence, "peacekeeping" is required, with attempts to maintain peace and security, sometimes employing mixed groups of military and civilian personnel. Peacebuilding, therefore, is a more comprehensive process that includes and goes beyond peacemaking and peacekeeping, emphasizing the need for long-term peace through implementing beneficial reforms at the grassroots level (Lederach 1997).

Civil Society in Peacebuilding. After developing a grasp of civil society and peacebuilding, it is necessary to look into the interconnection between civil society and the multifaceted idea of peacebuilding. The recognition of the effectiveness of civil society is evidenced by the high utilization of both International and local governments as partners in peacebuilding (Donais 2012; Mac Ginty 2011). The literature further shows that civil society participation in peacebuilding is beneficial because it strengthens the durability of peace and develops bridging social capital. (Osamba 2001; Nilsson 2012; Krause et al. 2018). Civil society organizations have been increasingly active in conflict prevention, peacemaking, and peacebuilding initiatives (Barnes 2005; Paffenholz and Spurk 2006). A functional approach has been developed (Paffenholz and Spurk 2006), identifying various roles for civil society, such as protection, monitoring/early warning, advocacy/public communication, socialization, social cohesion, intermediation/facilitation, and service provision. These functions provide a foundation for understanding civil society's impact on peacebuilding efforts, ensuring citizen safety, holding conflict actors accountable, raising public awareness, promoting a culture of peace, and reducing inter-group violence. As support to States weakened by conflict and violence, civil society can offer citizens protection through programs that protect residents' lives, property, and rights during and after conflict—the Philippines' peace zones are an example

(Orjuela 2003). Monitoring is the action of civil society organizations, whether international or local, that holds conflict actors accountable (especially in executing the provisions of negotiations/agreements) while also giving information for early warning of a conflict situation. Advocacy is critical for local civil society in peacebuilding (Paffenholz 2010). Being community spokespeople, advocating and articulating the interests and demands of their members is critical to raising public awareness and gaining support for these demands and interests to be presented on the peace agenda. The purpose of socialization in peacebuilding is to promote a culture of peace. Socialization aims to instill an attitude change in many individuals as part of their efforts toward "peaceful conflict resolution and reconciliation" (Paffenholz and Reychler 2007, p. 18). Meanwhile, the social cohesiveness function seeks to "bridge social capital that can serve to reduce inter-group violence and renew group relationships, interdependence, and solidarity" (ibid:20). Meanwhile, the goal of intermediation and facilitation is to bring together (formal/informal) armed groups or conflict parties in order to establish an agreement. An informal discussion process (started by civil society) to establish trust can sometimes contribute to a better outcome of formal peace negotiations (Paffenholz 2010). Finally, when organizations employ service delivery as an entry point into peacebuilding, it can have a good impact. When services are offered efficiently to exclude people at the base of conflict mainly, the risks of violence escalation are reduced (ibid).

Building upon the recognition of civil society's positive role in peacebuilding, the current shift towards a "local turn" represents a departure from the Western-centric liberal peace agenda of the 1990s, which focused on Western perspectives and methodologies (Ojendal et al. 2017). The recognition of civil society's positive role in peacebuilding can be traced back to Lederach (1997; Paffenholz 2015), who considered civil society involvement in peace initiatives at the local and grassroots levels to be more likely to produce positive results than interventions initiated by external peacebuilders. Local peacebuilding is seen as an emancipatory strategy, avoiding the exclusion of the locals from decision-making processes. Structural conditions supporting local civil society effectiveness include decentralization, local government support for peace, strengthened local capacity, and local ownership (Verkoren and Leeuwen 2013; Paffenholz 2015; Leonardson and Rudd 2015). Localized engagements and prioritized civic actions focus on deeply understanding the local context and cultural dynamics and customized solutions to address community-specific concerns. This viewpoint emphasizes that effective peacebuilding necessitates a thorough understanding of the local environment and the specific context and circumstances (Lederach 1997; Galtung 1996).

1.2 Locating Bangsamoro Civil Society in Peacebuilding

Mindanao civil society organizations (CSOs) expanded after the People Power Revolt in 1986. Bangsamoro-led organizations emerged following the GRP-MNLF negotiations, leading to increased international and local funding for peacebuilding activities. This growth was driven by the GRP-MNLF peace deal, a milestone for early civil society engagement (Askandar and Abubakar 2009). Pres. Arroyo's administration introduced a comprehensive peace process, emphasizing community-based approaches and active citizen participation (Evangelista in Askandar and Abubakar 2009). Many CSOs participated in nationwide consultations, exploring the roots of conflict and suggesting solutions. The Office of the Presidential Adviser on the Peace Process (OPAPP) led the Peace Infrastructure initiative, partnering with various NGOs, POs, religious schools, and local government units to promote a culture of peace in the region. After the GRP-MNLF agreement, local service providers became crucial for implementing peace and development projects in MNLF areas. CSOs took on more significant roles in peacebuilding, monitoring ceasefire implementation, and other activities. A notable example is the region-wide Week of Peace in 1999, which showcased the potential for increased ownership through collaboration between diverse actors in Mindanao (McCann 2000).

The Memorandum of Agreement on Ancestral Domain (MOA-AD) peace negotiations' lack of success revealed the schisms present within civil society and their restricted impact on the peace process, as noted by Alim (2016) and Arnado (2009). To overcome these limitations, it was apparent that some Bangsamoro CSOs C necessitated a more profound comprehension of the area's peace and security concerns to augment their endeavors' efficiency, as Alim (2000) and Mastura in Arnado (2009) noted. Notwithstanding these obstacles, diverse endeavors to foster peace endured, such as advocating for a peaceful culture and coordinating peace-related gatherings (Rood 2005). The active involvement of the Bangsamoro Civil Society in the Civilian Protection Component and the International Contact Group has resulted in the establishment of a "hybrid" peace infrastructure, as highlighted by Alim (2016), underscores their significant contribution to the peacebuilding process. Women's participation in the peace process has also expanded, with female civil society leaders joining the negotiating panel and becoming advisers on the peace process (Busran-Lao no date).

The number of civil society organizations (CSOs) in the Bangsamoro region of the southern Philippines has increased dramatically. These CSOs, which include non-governmental organizations, indigenous peoples' organizations, women's organizations, faith-based institutions, and academic institutions, have emerged as significant actors in tackling the region's long-standing conflict, poverty, and inequality. Their diversified activities and services are rooted in a community-based approach and reflect their substantial program implementation expertise; with many CSOs having been in operation for over a decade, they have gained significant experience and expertise in program implementation.

1.3 Understanding the Bangsamoro and the Bangsamoro Conflict

"Bangsamoro" refers to a diverse group in the southern Philippines with a complex history and identity. The region is multi-ethnic and multi-lingual and has experienced a decades-long conflict with roots tracing back to the 16th century. The Bangsamoro struggle began as resistance against Spanish colonization, with the formation of the Maguindanao and Sulu Sultanates before the arrival of the Spaniards. The term "Moro" was used by the Spaniards as a derogatory label for the unconquered people of the South (Majul 1973). The American colonial power later subdued the region and integrated it into their new colony, causing further resentment and resistance. In the face of continued dominance, Moro representatives sought to reshape their identity and declared themselves "Muslim Filipinos" to advance their interests in the newly independent Filipino nation. However, some factions preferred the term "Moro" to evoke a sense of shared history and resistance (Caballero 2007). The Moro National Liberation Front (MNLF) and the Moro Islamic Liberation Front (MILF) became the most prominent armed groups representing the Muslim-Moro minority (Buendia 2005; Shiavo-Campo and Judd 2005). The conflict has evolved, with new violent armed groups, such as Abu Sayyaf and the Maute group. The prolonged conflict has deep roots in history, poverty, and inequality. Landlessness and the competition for resources have contributed to the high poverty level among the indigenous Moros.

Additionally, traditional leaders have often colluded with colonial elites, further exacerbating the situation (Vellema et al. 2011). The rise in violence and the emergence of more violent groups emphasize the need for a better response to address the complex factors driving the conflict. The Philippines is listed as the 12th deadliest country regarding terrorism, with Mindanao being the most affected area. The lawlessness of the ARMM region allows illicit economic activities to thrive, contributing to the persistence of conflict. Weak institutions, a flawed political system, and the inability to deliver basic services have also fueled the violence. In conclusion, understanding the Bangsamoro and the Bangsamoro conflict requires a comprehensive approach considering the local context, power dynamics, wealth distribution, and other social factors. Peacebuilding initiatives should address both the traditional grievances and the complex drivers of the conflict, including shadow economies and ethnic divides.

1.4 A Framework for Understanding the Contributions, Opportunities, and Challenges of Bangsamoro Youth Civil Society Organizations in Peacebuilding

This research employs a modified version of Paffenholz and Spurk's (2006) comprehensive analytical framework to comprehend the Bangsamoro youth civil society's contribution to peacebuilding which centers on two stages. Firstly, it involves identifying and assessing the activities of civil society in peacebuilding. Secondly, it analyzes the institutional factors within the particular context of peacebuilding utilizing Douglas North's institutional theory.

The seven civil society functions of peacebuilding identified by Paffenholz and Spurk (2006) include: 1. **Protection**: The protection function is a core function of democracy theory, dating back to John Locke and others. It ensures citizens and communities are protected against the despotism of the state and any armed actor, including the national army and local groups. This function is often attributed to external NGOs that support national or local civil society actors, either indirectly or directly. 2. **Monitoring**: is a precondition for the protection and advocacy/public communication functions. It is also an essential function in democratization to hold governments accountable. The monitoring function is based on Montesquieu's separation of powers and is enhanced by development-cooperation perspectives. 3. **Advocacy and Public Communication**: Advocacy, a core function within democracy discourse, entails civil society promoting relevant social and political themes on the public agenda. The main activities within this function include agenda-setting by local civil society actors, lobbying for civil society involvement in peace negotiations, and creating public pressure. Public communication focuses on claims and demands made in public via demonstrations, press releases, petitions, or other statements supporting a specific demand. Civil society plays a crucial role in peacebuilding and promoting democratic behavior. It can be linked to the official negotiation process through information campaigns, public opinion polls, and direct involvement. During armed conflict, civil society can advocate for achieving a peace agreement against violence and human rights violations and broad-based participation in the peace process. 4. **In-group socialization** is a crucial function that supports democratic behavior and upholds democratic attitudes and values within society. It aims to inculcate a "culture of peace" within divided societies, promoting attitude changes and developing peaceful conflict resolution and reconciliation. In-group socialization can be divided into two types: the culture of peace, which aims to enhance a culture of peace and constructive conflict resolution for society at large, or single groups within society or to the conflict, and the building or consolidating in-group identity, which aims to strengthen the identity of a particular group, most oppressed or marginalized groups in asymmetric conflicts. However, the current practice of in-group socialization shows weaknesses, such as being too sporadic, lacking coordination, and failing to create a critical mass movement for change. 5. **Social cohesion** is an essential civil society function that ensures community-building. Building "bridging ties" across adversarial groups is essential, not just bonding ties within specific groups. There are three social cohesion-oriented activities: relationship-oriented cohesion for peace, outcome-oriented cohesion for peace, and outcome-oriented cohesion for business or development work (non-peace). These activities help groups learn how to live together in peaceful coexistence and help rebuild trust and social capital. Ethnically integrated organizations, such as businesses, trade, and associations, have effectively built ties across ethnically divided groups and facilitated the control of violence. 6. **The intermediation function** of civil society within democracy discourse highlights the role of civil society as an intermediator/facilitator between citizens and the state. In the peacebuilding context, facilitation can occur between or among groups at different levels of society. The main activities within this function include facilitation initiatives between armed groups, armed groups, and communities and among armed groups, communities, and development agencies. Local civil society can often facilitate between civil society and warring parties on the village or district level, between warring parties to negotiate peace zones or violence-free days, between international or national aid agencies and warring parties, and between international or national aid agencies and local civil society. 7. **Service Delivery is the seventh function**. As state structures are destroyed or damaged during armed conflicts, civil society actors' provision of aid and services (primarily NGOs, associations, and faith-based organizations) grows. There is little doubt that this is critical to assisting war-affected populations. We frequently see the same actors offering services and other peacebuilding functions simultaneously.

Institutional Theory: The current investigation employs an institutional theory to investigate the factors that impact the involvement of Bangsamoro youth civil society organizations in peacebuilding. The study underscores the crucial role of institutions in shaping their undertakings within the BARMM's socio-cultural, political, and economic milieus.

The central claim of the proponents of Institutionalism is that 'institutions matter' (Peters 2005; Schmidt 2006), particularly by acting as a mediator in shaping the actor's conduct and the subsequent political outcomes (March and Olsen 1998; Hay et al. 2006). These scholars believe that focusing on Institutions rather than individuals is an excellent analytical tool. Institutionalists differ in how they define institutions and how institutions matter. Several definitions may be provided. One that may capture a general and comprehensive view of this study is provided by Douglas North (1990), who defines institutions as "the humanly devised constraints that shape human interaction" or "the rules of the game in a society." This formulation implies several crucial elements. First, institutions are built by humans to control behavior and influence incentives. Second, institutions contain a wide range of de facto and de jure characteristics, including economic, political, and social dimensions. Institutions are essentially political since different sets of institutions result in various resource allocations. North emphasizes the importance of institutions in shaping people's behavior by providing a framework of rules, norms, and expectations that influence their behaviors and relationships. This perspective emphasizes the importance of institutions in determining diverse societal outcomes, whether social, economic, or political, by affecting individuals' incentives and choices within a specific environment.

Based on Douglas North's institutional theory, this approach examines the interplay of sociopolitical, cultural, economic, formal, and informal institutional factors that drive the achievements and restrictions experienced by Bangsamoro Civil Society Organizations (CSOs) in their peacebuilding activities by taking into account these varied institutional elements. This approach attempts to shed light on the essential role of institutions in defining and facilitating civil society's contributions to peacebuilding in the Bangsamoro region by analyzing the intricate interconnections within various institutional realms.

The sociopolitical context influences the peacebuilding activities of Bangsamoro Civil Society Organizations (CSOs). It entails investigating the political structures, power dynamics, governance systems, and policies that determine the role and influence of civil society organizations (CSOs) in peacebuilding. This component investigates how CSOs' obligations and possibilities to participate in peacebuilding processes are shaped by legal and regulatory frameworks, political ideologies, and interactions with government institutions.

The cultural environment is another critical institutional factor influencing CSOs' efforts in peacebuilding. Understanding the cultural norms, beliefs, traditions, and identities of the communities with which they work is critical for successful peacebuilding efforts. This section investigates how cultural factors influence CSO' attitudes, behaviors, and relationships with the communities they serve. It investigates how civil society organizations (CSOs) adapt their peacebuilding strategies, messaging, and community involvement approaches to line with cultural values and sensitivities.

Economic context also substantially impacts CSOs' attempts to promote peace in the Bangsamoro region. This component examines the community's economic situation, inequities, resource distribution, and livelihood opportunities. It looks into how economic issues affect CSOs' capacity, resource mobilization, and long-term viability in executing peacebuilding efforts. It also investigates CSOs' economic issues and ideas for addressing economic inequities, promoting sustainable livelihoods, and creating economic possibilities as part of their peacebuilding efforts.

To understand and evaluate the role of the cases in this study, their activities, efficacy, and the factors influencing their efforts will be organized in three steps: First, Identifying activities and examining CSO peacebuilding efforts and functions by employing Paffenholz and Spurk's (2006) framework to comprehend their contributions to peacebuilding. Second, assessing the effectiveness of CSO operations in attaining peacebuilding goals such as reducing violence, negotiating agreements, and fostering conflict resolution. Finally, considering the specific context and drawing institutional theory, the study explores the sociopolitical, cultural, and economic elements that affect Bangsamoro CSOs' peacebuilding efforts.

2 Methodology

The research design employed a qualitative methodology to describe and analyze Bangsamoro youth civil society organizations' presence, functions, and contributions to Mindanao peacebuilding. A case study approach explored diverse youth organizations and the multifaceted socio-cultural and political backdrop that may have influenced their effectiveness and contribution to peacebuilding, thereby improving understanding of their success and yielding insights that can guide future peacebuilding efforts in Muslim Mindanao's Bangsamoro Autonomous Region. The selection of cases for this study was conducted through purposive sampling to guarantee a varied representation of Bangsamoro youth organizations involved in peacebuilding. The study sample comprised mature groups that have been in operation for over a decade, sourced from the Consortium of Bangsamoro Civil Society groups, and newly established organizations that have existed for five years or less, sourced from the Ministry of Peace, Order, and Security list. Due to the constraints imposed by the pandemic, various organizations were selected through a random invitation process to participate. The study included representatives who responded and conveyed their eagerness to participate while accommodating their schedules to ensure their inclusion in the research. Adopting this methodology was deemed imperative owing to the practical limitations presented by the pandemic and the unreliable internet connectivity, underscoring the significance of flexibility and responsiveness in the selection process.

Included in this study are United Youth for Peace and Development (UNYPAD), The Moropreneur Inc, Thuma Ko Kapagingud Service Organization (THUMA Inc.), Aretes Style, Maguindanao Youth Alliance (MAYA), Youth Development for Patikul Association, United Group of Active Youth Association Inc. (UNGAYAN). The study employed Key Informant Interviews (KIIs) and Focus Group Discussions (FGDs) to gather data from a heterogeneous sample of participants, comprising the head or member representative of the CSO organizations, government officials, academics, and community stakeholders who had previously undergone peacebuilding interventions.

The respondents were questioned using a semi-structured interview guide, which allowed for the collection of descriptions of the 'peacebuilding' environment in which the interviewee lives. Because of the pandemic, most in-depth interviews were conducted online. Zoom, G Meet, and phone interviews are all options. FGDs were also made in the area were the participants are available.

Moreover, this study also incorporated secondary supplementary sources, including official publications, surveys, reports, prior research, individual records, video materials and promotional presentations, and social media content. The data collected from various sources were subjected to thorough analysis using the QDA Miner, a software for qualitative data analysis, as well as manual coding techniques. Analyzing the data comprised several stages, namely data organization, refinement, categorization, and validation, which collectively enabled an in-depth knowledge of the research topic.

The study placed a high value on ethical issues (Hoglund and Orjuela 2011), emphasizing the importance of informed consent and community engagement. The study prioritized the safety of the participants by getting informed consent from all parties involved. The informed consent process entailed a thorough description of the study's aims, methodologies, as well as the potential risks and advantages of participation. The study ensured that participants' confidentiality and identities were protected, and they were free to withdraw at any time.

3 The Transformative Landscape of Bangsamoro Youth Organizations: Empowering Voices and Strategies for Peacebuilding

The nature and character of Bangsamoro youth organizations have experienced significant growth and transformation in recent years. Once only a small fraction of the broader civil society, the number of registered youth organizations in the Bangsamoro region has surged from 307 in 2020 to 536 today, spanning all five provinces and various Moro communities. This increase in representation has provided Bangsamoro youth with a more prominent voice and allowed them to participate in development programs and influence policy recommendations actively. The Bangsamoro Youth Commission (BYC) plays a crucial role in accrediting and supporting youth organizations, emphasizing the strength and dynamism of the young Bangsamoro population. These organizations can now participate and address issues affecting them through a parliamentary setup, contributing to a more transparent and inclusive democratic space. While some long-standing youth organizations, such as United Youth for Peace and Development, have existed for over a decade, many newer organizations have unique origins and motivations. Some were formed in response to the adverse effects of conflict on young people, while others emerged from educational and leadership experiences that inspired activism and community engagement. For instance, organizations like MAYA and UNGAYAN were founded by members who gained valuable exposure and training in high school. This exposure fueled their passion for activism and led them to join or create youth organizations. On the other hand, THUMA Inc. was established by friends who volunteered to distribute aid during the Marawi siege. This harrowing experience taught them a valuable lesson and inspired them to form their organization. The Bangsamoro youth organizations' expanding number and diverse nature have significantly impacted the region's civil society landscape. These organizations provide an essential platform for young people to influence change and contribute to the growth of a more inclusive and democratic society.

The Bangsamoro youth civil society has been working hard to position and strategize itself within peacebuilding. Considering the historical backdrop and the changing political landscape, these youth organizations have used various tactics to engage with various stakeholders and address their varied difficulties. Building strong alliances and collaborations with local and international actors is a fundamental tactic Bangsamoro youth organizations use. This includes joining with organizations such as UNYPAD, UNYPHIL-Women, and other older youth organizations that have previously supported peace talks between the Philippine government and the Moro Islamic Liberation Front (MILF). Youth organizations can use these coalitions to elevate their voices and influence public opinion on peacebuilding efforts. Maintaining political independence is another critical component of the Bangsamoro youth organizations' strategy. While some groups have ties to the MILF or other political bodies, they realize the significance of maintaining their independence when addressing local needs and responding to the consequences of long-standing violent conflicts. This balance allows them to keep their watchdog role while seeking financial backing from ruling powers. Furthermore, despite potential hurdles, the Bangsamoro youth civil society has been working to interact with local government units (LGUs) and other stakeholders. Despite some less-than-ideal experiences, these organizations continue to advocate for more open communication channels, negotiation, and deliberation with the state. This contributes to a more cooperative and welcoming climate for peacebuilding. Finally, the Bangsamoro youth civic society recognizes the significance of adjusting to local conditions and embracing a hybrid approach to peacebuilding. This entails integrating top-down structures with bottom-up interaction to ensure that peacebuilding initiatives are grounded in reality while also understanding the potential limitations of the Western liberal concept of civil society.

3.1 Empowering Communities, Fostering Peace: The Comprehensive Contributions of Bangsamoro Civil Society Organizations to Peacebuilding and Sustainable Development

The southern Philippines' Bangsamoro region has a complicated history of conflict and instability. Various civil society organizations have emerged as beacons of hope in the face of these problems, working persistently to promote peace, social harmony, and sustainable development in their communities. Youth-led initiatives have played a critical role in pushing good change and developing a sense of unity among the younger generation. This section will present the contributions of youth organizations highlighting their work in pushing for human rights, gender equality, community development, and constructing a peaceful and inclusive society.

UNYPAD is a dynamic national organization dedicated to peacebuilding and community development. Their advocacy efforts are geared toward encouraging peace, togetherness, and solidarity among various communities. UNYPAD raises awareness and develops a conducive atmosphere for development and stability by participating in peace rallies and unity events. Their commitment to combating corruption improves their influence in developing a peaceful community. Community members are empowered by the organization's service delivery activities, particularly their skill and livelihood training programs. This enables individuals to produce income and better their socioeconomic circumstances, so addressing underlying causes that can contribute to conflict and instability. UNYPAD is critical in fostering long-term peace in the communities it serves by supporting socioeconomic development. One of UNYPAD's significant accomplishments is its role in conflict facilitation and mediation. They demonstrate their competence as intermediaries and peacebuilders by resolving long-standing feuds between powerful clans. UNYPAD assists communities in finding peaceful resolutions to issues, encouraging reconciliation and harmony by mobilizing elders, training them on conflict resolution approaches, and facilitating dialogue and consultations. UNYPAD contributes to peacebuilding by tackling socioeconomic challenges, empowering adolescents via skill training, supporting conflict resolution, and boosting community development. Their involvement in conflict resolution and service delivery contributes to reducing violence and establishing long-term peace in the communities they serve (Unypad 2020).

TheMoropreneur Inc. (TMI) is a social enterprise to empower the Bangsamoro and Indigenous Peoples (IPs) by co-creating innovative solutions for peaceful communities. The group supports creative solutions that match peacebuilding objectives and focuses on community-centered approaches. TMI's primary contributions to peacebuilding are capacity-building activities and instructional programs. This service improves the ability of individuals and communities to address conflicts and promote peace by providing them with knowledge, skills, and resources. Moreover, through social entrepreneurship, TMI has contributed to socioeconomic development. This was done by targeting core causes of conflict, including inequality and economic marginalization, and fostering economic growth, poverty reduction, and social well-being. Their initiatives help community stability and peace by encouraging inclusive development and giving economic opportunities. Further, the group is also engaged in advocacy and discourse. TMI increases awareness, impacts policies, and fosters collaboration through engaging in discussions with stakeholders, legislators, and community people. Their advocacy efforts foster peaceful cohabitation, inclusive development, and youth involvement, all contributing to the broader peacebuilding process. The comprehensive approach to peacebuilding used by The Moropreneur, which combines capacity-building, socioeconomic development, advocacy, and discourse, highlights the potential of social enterprises in tackling the issues encountered by conflict-affected communities. The Moropreneur contributes to peacebuilding by encouraging social cohesion, economic stability, and inclusive governance by concentrating on sustainable development and empowering marginalized communities.

THUMA Incorporated is a peacebuilding group that promotes community engagement and protection against violent threats. They place a premium on socialization and social cohesiveness, attempting to bridge the gap between state and non-state actors through innovative methods such as sports and peace education. One of their primary projects is to empower and educate young community development actors and moms about their role in peacebuilding. THUMA actively promotes social cohesion by bringing together socially excluded adolescents and organizing events to promote peace and relive the community's rich culture and tradition. They work with other groups, such as Save the Children Philippines and the Presidential Adviser on the Peace Process, to organize peace rallies and activities like the Marawi Week of Peace. Through conversations and training, THUMA enabled 30 Marawi City adolescents to contribute to social cohesion and become responsible players in averting violent extremism. THUMA engages in advocacy functions in addition to their focus on socialization and social cohesion. They provided gender-responsive capacity building, empowering 50 IDP and non-IDP Saguiaran (Municipality) women to actively avoid violent extremism within their families and communities. THUMA also implements service delivery programs, such as assisting with livelihoods, distributing personal safety equipment, and serving as a local counterpart in impact assessment conversations. THUMA Incorporated's multifaceted approach to peacebuilding has had a tremendous influence in its five years of existence, covering socialization, social cohesiveness, advocacy, and service delivery. THUMA supports peacebuilding efforts in their region of operation by encouraging understanding, trust, and tolerance among community members and tackling diverse societal concerns.

ARETES STYLE, a social enterprise established in 2019, was founded by a group of youth, including a chairperson who had experienced internal displacement. The primary objective of this innovative initiative is to co-produce alternative means of livelihood for the internally displaced women of Marawi City. By manufacturing hand-woven goods, Aretes Style aims to promote economic growth, preserve the rich Maranaw culture, and empower women in the process.

Aretes Style aids poverty reduction and economic empowerment by providing economic possibilities and increased INCOME for weavers. This socioeconomic development addresses the root causes of conflict and instability. Aretes Style has made significant strides in cultural preservation and dissemination. Aretes Style develops a sense of identity, pride, and a positive image for the Maranaw community by presenting Marawi City's distinct cultural heritage by producing hand-woven items. Cultural awareness and appreciation can promote social cohesion and peace. The organization's emphasis on helping internally displaced women reflects its dedication to gender equality and women's empowerment. Aretes Style encourages gender inclusiveness and empowerment through fair compensation and participation in decision-making processes. This can lead to better family dynamics, less domestic violence, and more women participating in decision-making processes, all of which are critical for long-term peace. Aretes Style's effort demonstrates the power of social entrepreneurs in solving economic, cultural, and gender issues in conflict-affected regions. Aretes Style aids peacebuilding by addressing core causes of conflict, promoting inclusivity, and supporting sustainable development by integrating economic development, cultural preservation, and women's empowerment.

The Maguindanao Alliance of Youth Advocates (MAYA) is a youth-led group that promotes women's rights, gender equality, and the abolition of child, early, and forced marriages. MAYA raises awareness and encourages good change through collaborations with religious leaders, influencers, and community members. Their lobbying aims to build a society where women and girls can express their rights, live in a safe environment, and access education and opportunity. MAYA actively involves community members through workshops, team-building events, and social activities, strengthening youth social cohesiveness. MAYA fosters a sense of unity and solidarity among young people by breaking down barriers and increasing understanding. This helps to establish peaceful and inclusive societies. The organization's impact on peacebuilding can be seen in its efforts to combat gender-based violence and promote the rights of women and children. MAYA contributes to creating an inclusive and equitable society in which everyone can live in peace and harmony through their advocacy and social cohesion projects. MAYA addresses core causes of conflict and establishes a more equitable and peaceful society by empowering women and promoting gender equality.

UNGAYAN (United Group of Active Youth Association Incorporated) advocates for youth human rights and encourages active engagement in community development. UNGAYAN promotes youth participation by providing training in human rights and parliamentary governance. Its goal is to establish an environment that respects and protects the rights of young people. Advocacy actions of the group include engaging legislators, community leaders, and stakeholders to increase awareness of youth concerns and promote young participation. UNGAYAN helps to create an enabling environment for youth development and peaceful coexistence by influencing policy and fostering collaboration. Through activities, workshops, and meetings, UNGAYAN promotes socializing among its members and other adolescents in the province. This contributes to their identity as a cohesive group working toward common aims. UNGAYAN supports peacebuilding by empowering young people and encouraging their active engagement in community development by strengthening their collective voice and advocating for their rights and interests.

Patikul Association's Youth Development organization in Sulu concentrates on peacebuilding efforts despite limited resources. The organization intends to repair the municipality's unfavorable reputation by involving youth in efforts to eradicate violence and encourage good change. They generally provide "culture of peace" training and workshops to socialize people. These activities strive to strengthen the organization's members' identities through a shared history, experiences, and ambitions. Through education and mentoring, the group tries to instill a better and more reasonable attitude among adolescents who may carry hostile and aggressive impulses. The Patikul Association's Youth Development has hurdles in changing the community's deep-seated hurt and hatred. Despite their attempts, changing unfavorable impressions and disassociating Patikul from violence is complex. Their emphasis on socializing and offering opportunities for youth to develop teamwork, leadership, trust, and solidarity, on the other hand, helps to form a cohesive community. The group contributes significantly to Patikul peacebuilding by addressing underlying issues and developing a culture of peace.

The initiatives of Bangsamoro civil society organizations contribute considerably to regional peacebuilding. These roles address traditional grievances and the conflict's more complex sources, such as poverty, inequality, and past injustices. These organizations promote inclusive development, social cohesion, and resilience in adversity by focusing on protection, monitoring, advocacy, socializing, intermediation, and service delivery. Their initiatives address current needs and lay the groundwork for long-term peace and stability in the Bangsamoro region.

3.2 *Unraveling Institutional Factors Shaping Engagement in the Bangsamoro Region: Empowering Youth for Peacebuilding*

Youth have emerged as formidable peacebuilding agents in the Bangsamoro region's changing terrain, making significant contributions across numerous sectors. This section delves into the complex web of formal and informal institutional variables that enable or inhibit young people and civil society organizations (CSOs) in their noble quest for peace. By diving into these institutional dynamics, we identify the critical characteristics that foster and support youth engagement, as well as the challenges and obstacles that impede their peacebuilding efforts.

Legal and Formal Institutional Frameworks for Peacebuilding. The Bangsamoro Youth Act, which went into effect on February 28, 2020, and ensures freedom of association and expression, is a crucial step toward institutionalizing young engagement in Bangsamoro. The law acknowledges the importance of youth in nation-building and encourages and protects their moral, spiritual, intellectual, and social well-being. The law also recognizes religion, faith, or one's traditional belief system as having a unique role in the holistic development of youth in the Bangsamoro region.

Individual young leaders and youth organizations have been able to engage with the Bangsamoro Youth Commission (BYC) and other BARMM agencies through collaboration or by providing venues for participation. The BYC's Bangsamoro Young Leaders Program has received P20,000 in project money from the Ayala Foundation and the Australian Embassy. The Ministry of Interior and Local Government (MILG) held a Training Workshop in Zamboanga City on the drafting of the Local Youth Development Plan (LYDP) to empower youth by imparting knowledge on various aspects of leadership and governance, as well as how these may be contextualized in the Bangsamoro society. The Ministry of Public Order and Safety offers various peacebuilding partnerships and initiatives for youth, including "peace promotion training workshops," which aim to engage young in peacebuilding, notably reconciliation, and unification. On the other hand, identifying youth as people aged 15–40 is a particularly significant component of the law, as it permits young professionals to communicate their concerns to the government.

The Peacebuilding Landscape: What rules? Local Turn, Liberal or Hybrid? The extensive use of international, national, and local governments as peacebuilding partners distinguishes the peacebuilding landscape in BARMM. The pivot of international peacebuilders toward local civil society organizations includes some form of local tun. However, most of the cases in this study show adhering to the templates and criteria of international and local engagement, showing that hybridity is more common than grassroots initiatives. The notion of development assistance serves as the foundation for local-international collaboration. Specific mechanisms are in place to provide help to local youth organizations. People in Need-Philippines cooperated with Thuma, Inc. through sub-granting, allowing civil society organizations or leaders from the informal sector to participate in their communities to undertake large or small-scale initiatives. Manaysay and Espesor (2020) underlined that in some local-outsider connections, local civil society organizations were given the power to develop their peace agenda an example would be the framework established by PIN in its support of THUMA's social cohesion project. Another case is that of Oxfam, the principal organizer and contributor to MAYA's participation in the "creating spaces project." The concept that the pooled efforts of various actors contribute to accomplishing global goals is the key guiding principle in collaborating with local civil society organizations. Oxfam's primary goal is to add value by bringing together different actors to work on common problems, supporting organizational and institutional capacity/strengthening, generating and sharing knowledge, promoting innovation and alternative solutions, and holding duty-bearers accountable for recognizing, protecting, and fulfilling women's and men's rights.

Development assistance and cooperation facilitate collaborations and interactions among many entities. However, many local civil society organizations suggest that in the context of international-local collaboration, local actors' agency and expertise should be more effectively leveraged. They claim that various external "support/assistance" programs frequently stymie the development of local organizations' ideas and activities because they do not consider local expertise and awareness of the situation. Furthermore, the Philippine government's stricter regulations limit the ability to execute the most recent norms and rules for providing help. These issues contribute to local civil society organizations' difficulties in fully leveraging their expertise and local context to contribute to effective peacebuilding efforts. Following the Marawi siege, there was a sense of distrust among Moro CSO, with funding and donations being misappropriated. As a result, a centralized CSO platform for the Marawi siege was established. However, it was led by a non-Moro entity. Turfing also occurred due to international donors' selection of local CSO partners, and newly created youth-led CSOs were denied the opportunity.

The data presented in this section point to unsuccessful peacebuilding efforts caused by some external donors' insensitivity to local conditions, culture, and capacities. For foreign interventions to be effective and long-lasting, "indigenous and insider methods" of conflict resolution must connect with and adapt to what local stakeholders accept and expect. A sensible option according to the institutionalism perspective, incentives from compliance with liberal peace conditionalities urge local actors or civil society to abide by the funders' terms. The Moro community knows that forming NGOs will provide them with financial security, but no one will openly accept it.

Political landscape. Civil society organizations have expressed concern over the region's political scene. Following the siege of Marawi, the Moro youth were more prone to indoctrination and recruitment and were labeled as terrorists. CSOs cannot accomplish their function of fighting and checking government abuses in the current political-security environment. In this context, CSOs can be bearers of truthful information that can assist in removing preconceptions against the people of BARMM. However, they are hampered from doing so because the situation is precarious. A respondent in Madalum, Lanao del Sur, emphasized the importance of the security sector in maintaining community peace and stability. However, some parents prefer not to chaperone their children because of the military presence, underlining the importance of solid civic-military connections in delivering peacebuilding initiatives.

Government institutions' role and norms in facilitating young activities are crucial. The present BARMM leadership initially opposed involving CSOs in government administration, but after several years of pushing "moral government," BARMM is now receptive to CSOs, particularly those of the youth. The findings of Adam et al. (2014) provide insight into the CSO-LGU relationship. The study found that when The Asia Foundation implemented a hybrid structure with UNYPAD as a partner for 'Community-level measures to strengthen local security in Mindanao,' political orders of the local state played a critical influence on the efficacy of interventions. There is a significant level of collaboration in Midsayap, Cotabato, with authoritative elite players monopolizing authority and flaunting a coercive form of government and socio-political control.

Most local administrations will only allow outside meddling if financial (monetary) and another tangible support is promised. Local administrations treat the community as if it were their fiefdom, and it is difficult to call them out. The worst practice is cooperation between local governments and civil society organizations, which makes it challenging to carry out an effective peacebuilding endeavor.

Socio-Cultural Context. The effectiveness of youth peace activists in the Bangsamoro region is heavily influenced by the socio-cultural setting in which they work. According to sociological institutionalism (SI), institutions are cultural norms, cognitive frames, and meaning systems that shape human behavior. In this setting, young women and members of youth civil society organizations frequently encounter helplessness in a patriarchal environment, as demonstrated by the experiences of organizations such as MAYA and Thuma, where age was viewed as a determinant of capability.

Despite the criticism of patriarchal elders, many young peace activists find courage in their Islamic beliefs, which motivates them to disregard abuse and severe insults. They are instilled with the appreciation of their critical role in pushing and questioning the Moro elders' standards in their community. However, these young women need help in influencing policy decision-making processes.

Bangsamoro society's social and ethnic-cultural distinctions significantly impact the performance and capabilities of emerging civil society organizations less felt by more mature CSOs. Despite these contrast, the younger organizations still actively engage due to their shared historical experiences with persecution and the devastation of war. Ethnic differences among Bangsamoro youth do not prevent them from participating in and contributing to peacebuilding efforts. However, CSOs from the island provinces have different opportunities than mainland ones.

One such harmful belief is twisted maratabat (honor), which has far-reaching consequences. Maratabat, a traditional value of Bangsamoro, is highly valued and firmly established in the community. When maratabat is misused or misinterpreted, it can harm the civil society of Bangsamoro youth and their peacebuilding efforts. It may contribute to a community's reluctance to identify and handle sensitive concerns or conflicts openly. This lack of open communication and conflict avoidance can stymie the resolution of underlying tensions and limit genuine discourse, both of which are required for effective peacebuilding. Furthermore, the notion of maratabat might limit the functioning of civil society organizations by framing them as family businesses or property. When civil society is mainly viewed as maximizing individual utility rather than collective action, it can stymie collaboration and attain common goals. Accordingly, this has impaired youth organizations' ability to contribute meaningfully to peacebuilding activities.

Understanding and resolving these socio-cultural variables is critical for empowering and supporting young Bangsamoro peace activists. We can promote a more vibrant and compelling youth-driven peacebuilding movement in the Bangsamoro region by navigating and questioning cultural norms, encouraging diversity, and creating spaces for youth voices to be heard and valued.

Networks and social capital are critical to the operation of fledgling civil society organizations. While development professionals' ideals should respect others, the relationships with other organizations might impact how a fledgling civil society group performs. Networks, especially seniors, should provide mentorship and assistance, yet a greedy mentality extends even to CSOs. Positive networks, such as CBCS, have been advantageous to youth CSOs by assisting them in embracing and owning their Bangsamoro identity and engaging the region's people. However, free-thinking youth have different experiences and principles than older and more mature CSOs, with whom they periodically fight. As a result, adolescents with opposing viewpoints are drowned out.

Being a part of a network gives you an advantage in a peacebuilding landscape that favors more established CSOs and networks. However, the network's politics drown out prospects for aid to 'unfavored' youth group members. The heads of the networks also run their non-profit organizations, which creates a problem with fairness.

Organizational values and capacities are also critical in the Bangsamoro young civic society. Most respondents stress positive individual qualities such as advocacy commitment. There needs to be a greater understanding of the idea of peacebuilding and the impact of local government processes on civil society engagement. Staffing for specific older youth organizations involves appointing family members, which can result in attrition and a low priority placed on wealth generation. This results in a failed CSO. The inability of youth CSOs to generate their funding, which is affected by membership fees and gifts from friends, is a typical issue. The shortage of accessible financial assistance is evident, and emerging youth CSOs have limited capacity to conduct initiatives and compete with established CSOs and consortium members, particularly on peacebuilding concerns.

4 Conclusion

Bangsamoro Civil Society Organizations (CSOs), particularly those led by youth, have made important contributions to peacebuilding efforts. These CSOs have played an important role in various sectors, collaborating with older CSO organizations in a complementary environment. However, the complexity of the conflict setting and a variety of institutional elements influence their efficiency in performing these functions. An examination of the peacebuilding activities of Bangsamoro civil society organizations indicates their active involvement in promoting peace and addressing conflict-related concerns. They have participated in efforts such as reconciliation, unification, and community development to restore trust and foster stability during the earlier periods of the peace process and through the aftermath of the Marawi siege. Given the difficulties and complexities of their conflict context, their efforts have been laudable. Regarding the institutional environment, there are both deterrent and enabling factors influencing the participation of youth CSOs in peacebuilding. On the one hand, there have been positive achievements, such as the passage of the Bangsamoro Youth Act, which recognizes and safeguards the role of youth in nation-building. Collaboration with government entities such as the Bangsamoro Youth Commission, the Ministry of Interior and Local Government, and the Ministry of Public Order and Safety has further increased participation and capacity-building possibilities. However, difficulties persist. Cultural conventions, patriarchal systems, and erroneous concepts of honor, such as maratabat, can all stymie the efficacy of youth CSOs in peacebuilding. The dominance of established CSOs and networks and limited access to financial resources create challenges for nascent youth-led groups. These issues underline the importance of creating a more welcoming and supportive atmosphere that recognizes the agency and skills of young peace activists. Exploiting the institutional component of inclusive governance and acknowledging the particular perspectives and skills of youth can help strengthen peacebuilding efforts. Empowering youth through meaningful engagement, supporting their ideas, and providing financial resources is critical. Fostering more significant partnerships across government institutions, civil society organizations, and youth organizations can open up areas for collaboration and coordination, allowing for more successful peacebuilding. The practical importance of these findings is that they can promote long-term peace and development in the Bangsamoro region. The area can establish a more inclusive and participatory peacebuilding framework by utilizing young peace activists' energy, ideas, and agency. The theoretical significance is that it sheds light on the interplay of cultural, institutional, and conflict elements, emphasizing the importance of context-specific methods to peacebuilding. However, difficulties still need to be solved in youth peacebuilding. CSO distrust, limited financial resources, and the intricacies of the conflict situation all provide substantial challenges. Practical steps can be taken to solve these difficulties. It is critical to strengthen ties between local and international players, provide financial support to youth CSOs, and promote intergenerational communication and collaboration. Raising awareness about the need for youth engagement and confronting harmful cultural attitudes can also help to create a more suitable atmosphere for youth peacebuilding. Further research into the dynamics of youth peacebuilding in the Bangsamoro region should focus on the role of institutional elements, cultural norms, and conflict context in their success. Exploring creative techniques and strategies to overcome obstacles and boost young participation might also be beneficial study areas. Finally, youth civil society organizations in the Bangsamoro region have significantly contributed to peacebuilding. Exploiting the institutional aspect of inclusive governance, eliminating cultural hurdles, and providing the appropriate support and resources are all critical stages in increasing its efficacy. The region can encourage long-term peace and growth by empowering young peace activists and establishing an enabling atmosphere.

References

Adam J, Verbrugge B, Vanden Boer D (2014) Peacemaking and state-society interaction in conflict-torn Mindanao, Philippines. The Justice and Security Research Programme and The Asia Foundation. Pdf

Alim G (2000) Civil society in Mindanao: way forward. In: Askandar, Kamarulzaman, Ayesah Abubakar (eds) 2009. Peace for Mindanao. SEACSN Publications, Penang. pp 118–205

Alim G (2016) Peacetalk: role of CSOs in peace building and political transition in Mindanao. Mindanews. Speech delivered by Guiamel Alim at the Global Autonomy, Governance, Federalism forum on Day 2, 20 October 2016 at Dusit Thani hotel in Makati City

Anheier H (2004) Civil society. Measurement Evaluation, Policy. Civicus. Earth Scan, London

Arnado MA (2009) In: Askandar, Abubakar (eds) Peace for Mindanao. SEACSN Publications, Penang

Askandar K, Abubakar A (eds) (2009) Peace for Mindanao. SEACSN Publications, Penang

Belloni R (2001) Civil society and peacebuilding in Bosnia and Herzegovina. J Peace Res 38(2):163–180

Barnes C (2005) Weaving the web: civil society roles in working with conflict and building peace. In: van Tongeren et al (eds) People building peace II: successful stories of civil society. Lynne Rienner Publishers, Boulder and London

Bercovitch J, Kadayifci A (2002) Exploring the relevance and contribution of mediation to peacebuilding. Peace Conflict Stud 9(2), Article 2. https://doi.org/10.46743/1082-7307/2002.1025

Bird L (2007) Learning about war and peace in the Great Lakes Region of Africa. Res Comp Int Educ. https://doi.org/10.2304/rcie.2007.2.3.176

Boege V (2011) Hybrid forms of peace and order on a South Sea Island: experiences from Bougainville (Papua New Guinea). In: Richmond OP, Mitchell A (eds) Hybrid forms of peace: from everyday agency to post-liberalism. Palgrave Macmillan, New York, pp 88–106

Buendia G (2005) The State-Moro armed conflict in the Philippines Unresolved national question or question of governance? Asian J Political Sci 13(1):109–138. https://doi.org/10.1080/02185370508434252

Busran-Lao YND. Philippine Women and Inclusivity in the Peace process. ACCORD

Caballero M (2007) Revisiting the Bangsamoro struggle: contested identities and elusive peace. Asian Secur #:2:141–161. https://doi.org/10.1080/14799850701351425

Campo S, Judd M (2005) The Mindanao conflict: roots, causes, and potential dividends. Social Development Papers: conflict Prevention and Reconstruction. https://www.worldbank.org. Accessed 9 Apr 2018

Donais T (2012) Peacebuilding and local ownership: post-conflict consensus-building. https://doi.org/10.4324/9780203118078.10.4314/AJCR.V2I1

Galtung J (1996) Peace by peaceful means: peace and conflict, Development and Civilization International Peace Research Institute, Oslo (PRIO)

Hampson F (1996) Nurturing peace: why peace processes succeed or fail. United States Institute of Peace, Washington, DC

Hay C, Lister M, Marsh D (2006) State theories and issues. Palgrave MacMillan, New York

Hoglund K, Orjuela C (2012) Hybrid Peace Governance and illiberal peacebuilding. Global Governance, pp 89–104

https://www.facebook.com/THUMA-1684720088493149. https://sites.google.com/view/thumalanao
https://www.facebook.com/search/top?q=maguindanao%20alliance%20of%20youth%20advocates
https://www.goldininstitute.org/news/9-leadership-development/889-bangsamoro-youth-in-promoting-peace-in-the-new-normal-updates-from-southern-philippines.
http://www.internationalalert.org/sites/default/files/publications/voices_from_the_villagescover.pdf
International Alert (2007) Voices from the villages: people's Agendas for peace
Krause J, Krause W, Bränfors P (2018) Women's participation in peace Negotiations and the durability of peace. Int Interact. https://doi.org/10.1080/03050629.2018.1492386
Kreuzer P (2011) Protracted civil war in Mindanao: can civil society help cut the Gordian knot? In: Quimpo N, Kasuya Y (eds) Politics of change in the Philippines. Anvil Publishing, Inc., Pasig City, pp 313–338
Lara F Jr, Champain P (2009) Inclusive peace in Mindanao: revisiting the dynamics of conflict and exclusion. International Alert, Quezon City
Lederach JP (1997) Building peace: sustainable reconciliation in divided societies. US Institute of Peace Press, Washington D.C
Leonardson H, Rudd G (2015) The 'local turn' in peacebuilding: a literature review of effective and emancipatory local peacebuilding. Third World Q 36(5):825–839. https://doi.org/10.1080/01436597.2015.1029905
Leeuwen M, Nindorera J, Nzweve J, Corbijn C (2020) The 'local turn' and notions of conflict and peacebuilding—Reflections on local peace committees in Burundi and eastern DR Congo. Peacebuilding. https://doi.org/10.1080/21647259.2019.1633760
March G, Olsen P (1998) The Institutional dynamics of international political orders. Int Organ 52(4):943–969
Mac Ginty R (2010) Hybrid peace: the interaction between top-down and bottom-up peace. Secur Dialog 41(4):391–412
Mac Ginty R (2011) International peacebuilding and local resistance: hybrid forms of peace. Palgrave Macmillan, Basingstoke, p 2011
Mac Ginty R, Richmond (2013) The local turn in peacebuilding: a critical agenda for peace. Third World Q 34(5):763–783
Majul C (1973) Muslims in the Philippines. UP Press, Quezon City
Manaysay F, Espesor J (2020) Frictional binaries: hybridity, civil society and liberal-local peacebuilding. In: Uesugi Y et al (eds) Mindanao in operationalisation of hybrid peacebuilding. Security, development and human rights in East Asia. Palgrave Macmillan, Cham. https://doi.org/10.1007/978-3-030-67758-9_6
McCan S (2000) Building a culture of peace amongst Muslims and Christians in Mindanao, Southern Philippines-the merging of theory and praxis. Int J Hum Peace 16(1):60. https://link.gale.com/apps/doc/A73580055/AONE?u=ussd&sid=AONE&xid=a455208f. Accessed 10 Sept 2019
Nilsson D (2012) Anchoring the peace: civil society actors in peace accords and durable peace. Int Interact. https://doi.org/10.1080/03050629.2012.659139
North D (1990) Institutions, institutional change, and economic performance are all important considerations. The Cambridge University Press
Osamba J (2001) Peace building and transformation from below: indigenous approaches to conflict resolution and reconciliation among the pastoral societies in the Borderlands of Eastern Africa. Afr J Conflict Resolut
Orjuela C (2003) Building peace in Sri Lanka: a role for civil society? J Peace Res 40(2):195–212. https://doi.org/10.1177/0022343303040002004
Ojendal J, Leonardsson H, Lundqvist M (2017) Local peacebuilding-challenges and opportunities. Stockholm: The Expert Group for Aid Studies (EBA) Report, 5.
Pafenholz T (2013) International peacebuilding goes local: analyzing Lederach's conflict transformation theory and its ambivalent encounter with 20 years of practice. Peacebuild J

Paffenholz T (2010) Civil society and peacebuilding: a critical assessment. Lynne Rienner Publishers, Boulder, USA
Paffenholz T (2015) Unpacking te local turn in peacebuilding: a critical assessment towards an agenda for future research. pp 857–874. https://doi.org/10.1080/01436597.2015.1029908
https://www.tandfonline.com/toc/rpcb20/2/1, https://doi.org/10.1080/21647259.2013.783257
Paffenholz T (2018) Unpacking the local turn in peacebuilding: a critical assessment towards an agenda for future research. (pp. 41–58). Routledge. https://doi.org/10.1080/01436597.2015.1029908
Paffenholz T, Spurk C (2006) Civil society, civic engagement, and peacebuilding. Soc Dev Pap 36
Paffenholz T, Reychler L (2007) Aid for peace: a guide to planning and evaluation for conflict zones
Peck C (1998) Sustainable peace: the role of the UN and regional organizations in preventing conflict. Rowman & Littlefield, London
Peters BG (2005) Institutional theory in political science: the new institutionalism. Ashford Colour Press, Gosport Hampshire. http://books.google.nl/books?hl=nl&lr=&id=HXV2esV. Accessed 08 June 2012
Richmond O, Franks J (2009) Liberal peace transitions: between statebuilding and peacebuilding. Edinburgh University Press
Rood S (2005) Forging sustainable peace in Mindanao: the role of civil society. Policy Stud 17 (East-West Center, Washington)
Rooy A (2004) The global legitimacy game: civil society, globalization and protest. Palgrave Macmillan, New York
Salamon LM (2010) Putting the civil society sector on the economic map of the world. Ann Publ Coop Econ 81(2):167–210. https://doi.org/10.1111/j.1467-8292.2010.00409.x
Steffek J, Nanz P (2007) Emergent patterns of civil society participation in global and European governance. In: Steffek J, Kissling C, Nanz P (eds) Civil society participation in European and global governance: a cure for the democratic deficit?. Palgrave Macmillan, New York, pp 1–29
Schmidt V (2006) Institutionalism. Chapter 5 In: Hay et al (eds) The state. Theories and issues. Palgrave MacMillan, New York, pp 98–117
Thuma Inc. https://sites.google.com/view/thumalanao/home. Accessed 2 Mar 2022
United Nations Security Council (2005) Resolution 1325. United Nations, New York
UNGAYAN. https://www.facebook.com/ungayadpm
UNYPAD. https://unypad.org/. Accessed 20 Feb 2020
Vellema S, Borras S Jr, Lara F Jr (2011) The agrarian roots of contemporary violent conflict in Mindanao, Southern Philippines. J Agrar Change 11:298–320
Verkoren W, van Leeuwen M (2013) Civil society in peacebuilding: global discourse local reality. Int Peacekeeping 20(2):159–172
Youth Development of Patikul. https://www.facebook.com/profile.php?id=100064694980870

Understanding the Role of the Melanesian Spreadhead Group (MSG) in the West Papuan Conflict

Aaron Denison Deivasagayam

Abstract The study of regional organisations has always revolved around the European Union (EU) or the Association of Southeast Asian Nations (ASEAN) which are frequently categorised as the two most successful models of regional integration often overshadowing other regional and sub-regional organisations despite the fact that these smaller organisations have had invaluable contributions to their respective regions. In this chapter, the author expands the field of research on regional and sub-regional organisations by highlighting the significant role of the Melanesian Spearhead Group (MSG), a sub-regional organization from the Pacific Islands in showcasing their contribution towards conflict management. Through the use of the preventive diplomacy framework, the author argues that despite being a relatively minor sub-regional organisation, the MSG has very much been instrumental by starring as a functional and credible actor in the West Papuan conflict. The findings highlight MSG's contribution and implementation of numerous preventive diplomacy strategies, which have resulted in constructive engagements between West Papua and Indonesia, preventing further escalations in addition to minimising the conflict between both parties despite outgoing challenges that has persisted in the province of West Papua.

Keywords West Papua · Melanesian Spearhead Group · Regionalism · Preventive diplomacy · Conflict

1 Introduction

Internal conflicts or disputes have a wide range of consequences for specific states, particularly in terms of economic development and security. States embroiled in violent conflicts tend to make slower progress in terms of development and economic growth due to ongoing conflicts that hinder progress (Kim 2006). States in conflict

A. D. Deivasagayam (✉)
Graduate School of Humanities and Social Sciences, Hiroshima University, Hiroshima, Japan
e-mail: d225511@hiroshima-u.ac.jp; aaron16denison@gmail.com

A. Sharifi et al. (eds.), *Bridging Peace and Sustainability Amidst Global Transformations*, World Sustainability Series,
https://doi.org/10.1007/978-981-99-7572-3_7

are vulnerable as they prioritize security over development in their national budgets, resulting in low economic growth and poverty. The internal conflicts in Pattani (Southern Thailand) and Mindanao (Southern Philippines), are two excellent examples that very much resonates with the vulnerabilities mentioned above, with both internal conflicts having dominated the Southeast Asian region for many years now (Vatikiotis 2009) In Thailand, the government is believed to have spent almost USD 6.1 billion on defence in 2017, with the figure rising to USD 7.3 billion in 2019, with a large portion of this going to the conflict-affected region of Pattani in southern Thailand (Parameswaran 2016; ASEAN Today 2021). Similarly in the Philippines where military forces have been heavily concentrated or stationed in the conflict-prone region of Mindanao in Southern Philippines as a precaution for national security (Abuza 2019).

In addition to the internal conflicts in Pattani and Mindanao, another conflict in Southeast Asia that has received little attention is the internal conflict in the province of West Papua in Indonesia. The West Papuan conflict dates back to the late 1940s, with an initial power struggle between Indonesia and the Netherlands over the region, which eventually turned into a power struggle between Indonesia and the people of West Papua via the Act of Free Choice 1969, a referendum promised to the people of West Papua (Philpott 2018). The referendum became a source of contention as it was deemed to have not represented the majority of the population in West Papua at the time with pro-independence groups in West Papua challenging the overall outcome of the referendum. This is due to the fact that only 1025 West Papuans out of the overall population of 800,000 were allowed to vote in the referendum and all those who voted, overwhelmingly desired to remain part of Indonesia (Scott and Tebay 2005). Despite opposition in the province, the Indonesian government has remained firm in its endorsement for the referendum's outcome. The dispute between the Indonesian administration and local West Papuans persists to this day, morphing into an internal conflict within the province. While it appears that Indonesia has everything under control, the sub-regional organisation of the Melanesian Spearhead Group (MSG) has played a decisive role in avoiding further escalations between both parties through its regional presence. The role of the MSG as a sub-regional organization in this conflict has been crucial as there has been evidence of various contributions and efforts undertaken by the MSG through preventive diplomacy strategies in attempting to mitigate the conflict between West Papua and Indonesia.

Thus, the aim of this chapter is to analyse MSG's role as a sub-regional organisation in managing the West Papuan conflict. The main argument of the chapter highlights MSG's influence in the West Papuan conflict through the many different contributions and efforts in ensuring continuous dialogues between both West Papua and Indonesia as a way of minimising the conflict, which hopefully leads to a resolution in the future. The chapter begins with an overview of the preventive diplomacy framework as well as the different methods and mechanisms of implementing them under various circumstances or case studies. The following section of the chapter explores the background and current developments of the conflict in West Papua as well as the history of the MSG as a sub-regional organisation. The final section provides an analysis of the MSG's role in managing the West Papuan conflict through the preventive diplomacy framework, highlighting the various preventive diplomacy strategies utilised by the MSG which contributed directly to the minimisation of the West Papuan conflict, despite the ongoing challenges in the province.

2 Conceptualising the Preventive Diplomacy Framework

In this chapter, MSG's role in the West Papuan conflict is examined through the preventive diplomacy framework which is an important concept in the area of conflict management. The initial conceptualisation of preventive diplomacy emerged from Dag Hammarskjöld, who was known as the *dove of preventive diplomacy* during his tenure as the second United Nations (UN) Secretary General from 1953 to 1961 (Fröhlich 2018). According to Hammarskjold, the goal of preventive diplomacy was to basically make sure that internal or local conflicts are limited within a small scope by avoiding a full-scale war among global powers (Knight and Yamashita 1993). The function of preventive diplomacy was believed to have been neglected for nearly forty years after Hammarskjold's death before being revitalized by the sixth UN Secretary General, Boutros Boutros-Ghali through his report titled An Agenda for Peace in 1992. In the report, Boutros-Ghali (1992) defines preventive diplomacy as an act of preventing disputes between two or more parties as well as preventing current disputes from escalating further into becoming conflicts in addition to controlling the spread of a conflict in case it eventually happens or even escalates. Along the years many scholars have expanded and developed the notion of preventive diplomacy in accordance with the UN based on the latest happenings around the world especially in responding to new challenges globally (Jenča 2013). According to Lund (2001), preventive diplomacy is viewed as an action taken by states in vulnerable areas to avoid the deployment of armed forces and related forms of coercions to settle disputes. Ozcelik (2006) notes that the most efficient form of preventive diplomacy is the ability to resolve hostilities before they develop into a conflict and to respond instantly in minimizing a conflict when it eventually breaks out. While there are numerous conceptualisations of preventive diplomacy, most of them basically tend to have similar keywords or ideas such as preventing larger conflicts, minimising conflicts before escalations and the avoidance of using force. In terms of the actors of preventive diplomacy, most of the initiatives at least in the initial stages was very much centred around and led by the UN, however the eighth UN Secretary General, Ban Ki-Moon in 2011 acknowledged the importance of other agencies or organisations as important actors in the development of preventive diplomacy in line with the evolvement of the concept itself (Moon, 2011). Alongside the UN, which remained as an influential actor, individual states especially middle power states became important actors in taking up the role and responsibilities of advancing preventive diplomacy (Cordovez 1987). In addition to individual states, even new networks of NGOs as well as regional and local community-based organisation emerged as significant actors in addressing preventive diplomacy (Muggah and White 2013). Crucially, even regional and sub-regional organisations also began engaging in preventive diplomacy both independently as well as in partnership with the UN (Mancini 2011; Zyck and Muggah 2012). Thus, the actors of preventive diplomacy has developed beyond the UN with the inclusion of local and regional civil society organisations as well as regional and sub-regional organisations that has very much become instrumental in addressing conflicts.

Upon comprehending the capabilities of various actors in fulfilling their role through preventive diplomacy, the next step is to understand the different strategies or approaches in carrying out preventive diplomacy. Some of the initial approaches carried out by the UN in implementing preventive diplomacy include trust building, early warning, preventive deployment, establishment of demilitarized zones as well as peacekeeping operations (Boutros-Ghali 1992; Yan and Guo 2020). As preventive diplomacy actors expanded beyond the UN, regional and sub-regional organisation began implementing their own preventive diplomacy strategies in managing conflicts and these strategies were mostly extensions of those carried out by the UN. Among some of the methods or strategies that were often used at least at the regional level in managing conflicts were negotiations that were led by the municipal or city (Cohen et al. 2010), councils of the wise (Murithi and Mwaura 2010), fact finding (Borda 2011), formal mediation (Cristescu et al. 2012), good offices (Della-Giacoma 2011) and legal frameworks (Emmers and Tan 2009). Additionally, Babbitt (2012) also points out the utilisation of strategies such as the use of norms and rules, quiet diplomacy as well as election monitoring in carrying out preventive diplomacy. Drawing on the available approaches utilised by regional and sub-regional organisations in implementing preventive diplomacy, this chapter identified five strategies that are helpful in analysing MSG's role in the West Papuan conflict. For the benefit of the readers, I will be outlining the details of the five strategies that were identified and provide illustrative instances of how other regional and sub-regional organisations applied these strategies in the context of preventive diplomacy. The first strategy is the utilisation of the council of the wise, which serves in the modern world as an alternative for a cultural council that weighs in to arbritrate conflicts. Occasionally, the council is made up of a group of individuals that serves as mediators by providing assistance for capacity building initiatives in addition to sharing appropriate lessons that could be useful for the conflicting parties in an effort to prevent further escalation of conflicts (Murithi and Mwaura 2010; Cristescu et al. 2012). The African Union (AU) is very much familiar with this approach through its Panel of the Wise (PoW) which was established to contribute to the development of peace and security particularly in averting internal and external conflicts among its member states (Porto and Ngandu 2015). The AU's role in managing the crisis in the Central African Republic in 2017 very much demonstrated the role of the PoW which held consultations with political parties as well as armed insurgency groups as a way of diffusing tensions in the state that had reached a boiling point. Subsequently, the PoW submitted a report to the President of the Central African Republic at that time, François Bozizé recommending a continuous national dialogue between the different interest groups which ultimately resulted in the de-escalation of hostilities in the state (Ngandu 2017).

The second strategy is the use of norms and rules as leverage in preventing conflicts. According to Babbitt (2012) the objective of this strategy is to influence governments' preferences by encouraging compliance with certain standards or norms as a way of gaining international or regional approval for increased access to political and economic opportunities. These norms-based frameworks have occasionally been utilized as a shield by governments in implementing difficult recommendations. The High Commissioner on National Minorities (HCNM) of the Organization for Security and Co-operation in Europe (OSCE) demonstrated the use of norms-based framework especially in the translation of norms into practical guidance and concrete proposals for the consideration of minorities and governments such as the drafting of the language legislation Ukraine, the raising of money for Albanian universities in Macedonia and providing bilingual education for the minority Turks and Armenian in Georgia (Babbitt 2012). The compliance with certain norms as a common ground along with practical proposals by the HCNM prevented internal de-escalations in those three states by framing the initiatives in the three states mentioned above as in line with the objectives and standards practiced by the European community as way of attaining regional approval (Kachuyevski 2012). The third strategy is formal mediation, which in some cases are carried out by special envoys who in general, have a profound understanding of local sensitivities and have a strong rapport with either the local communities or the conflicting parties. The strategy of formal mediation was demonstrated in Guinea in the context of the controversial stadium massacre in 2009. During the 2009 crisis, the President of the Economic Community of West African States (ECOWAS), Umaru Musa Yar'Adua established the International Contact Group on Guinea (ICG-G) by appointing special representatives who became involved in the formal mediation process. (Yabi 2010). In this case, the formal mediation mechanism with the appointment of special representatives was indeed an important step in the management of the crisis in Guinea which was eventually resolved by the ECOWAS through the ICG-G via mediation in collaboration with the African Union (AU) and the UN which provided technical expertise (Yabi 2010; Hara 2011).

The fourth strategy is fact-finding expeditions, which are usually necessary for the acquisition of significant details and information about the condition in a particular state which is subsequently utilised as evidence when deciding on the outcome of a specific issue or in this case a conflict. The Organization of American States (OAS) utilised this strategy to deal with the Colombia-Ecuador dispute by establishing a fact-finding expedition to investigate Colombia's violation of Ecuador's sovereignty through the OAS Resolution 930. The outcomes of the OAS fact-finding expedition were deemed significant in determining the final settlement of the issue, which effectively prevented a conflict between the two states (Borda 2011). The fifth strategy is the use of legal frameworks or agreements. This approach is in essence, a preventive diplomacy tool which focuses largely on treaties or legal documents signed and agreed upon by two or more entities with the commitment of avoiding de-escalation. International treaties or conventions are also considered as legal framework or agreements as they require states to comply with certain rules and regulations on specific issues. The Declaration on the Conduct of Parties in the South China Sea, signed between China and ASEAN in 2002 is an example of a regional legal framework, which expressed the commitments and desire of both parties in pursuing their territorial claims by peaceful means, and this legal framework has somewhat been effective in de-escalating tensions between China and the ASEAN claimant states (Emmers and Tan 2009).

The five strategies or approaches outlined above has been proven successful in the many aspects of preventive diplomacy such as the resolution of conflicts, the avoidance of the use of force, or even the minimization of existing conflicts with the important role of regional and sub-regional organisations acting independently or in collaboration with the UN. In the context of the West Papuan conflict, this chapter analyses MSG's role by examining the numerous contributions through the preventive diplomacy framework. It also evaluates whether the preventive diplomacy strategies utilised by the MSG were able to minimise the West Papuan conflict. Nonetheless, before delving deeper into the role of the MSG's in the West Papuan conflict, the next section of this chapter, will explore the background of the conflict in West Papua as well as the background of MSG as a sub-regional organisation. It is critical that readers have a fundamental grasp of the conflict as well as knowledge of the development and evolution of the MSG as a sub-regional organization in order to have a deeper understanding of the linkages between the three important actors in this conflict namely West Papua, Indonesia and the MSG.

3 The Background of the Conflict in West Papua

West Papua is a province in the eastern part of Indonesia that borders the independent state of Papua New Guinea. The huge island is rich in natural resources such as gold, copper, timber and natural gases (Stott 2011). It is the largest province in Indonesia and was sometimes referred as Irian Jaya during the Suharto administration as it remained as such until 2000, before being renamed and seperated into the two provinces of Papua and West Papua in 2003. Nonetheless, the name West Papua is preferred by the locals throughout the province despite the administrative changes that occurred and this is very much due to the fact that majority of the original inhabitants in the province were Melanesian Papuans although in the national context majority identify themselves as Indonesians. The conflict in West Papua is not a new issue, it has been on the Southeast Asian radar since the late 1940s as a result of Indonesia's independence. The first point of contention of the conflict stemmed from a power struggle between Indonesia and the Netherlands. Indonesia had declared its independence alongside West Papua in 1945 however, when the Netherlands finally recognized Indonesia's independence in 1949, all provinces except West Papua were transferred to Indonesia (Lagerberg 1979). The Netherlands maintained that West Papua was administered as a separate territory due to the differences in identity of those who lived in the region thus it was not regarded as an Indonesian territory instead the Netherlands were preparing the region for its own independence (Lijphart 1966). The first Indonesian President post-independence, Sukarno demanded the incorporation of West Papua into their republic insisting that Indonesia was a nation incomplete without West Papua (Legge 2004).

Although the United States (US) and the United Kingdom (UK) initially supported the Netherlands' move to prepare West Papua for independence, Indonesia's strategic influence in relation to the Cold War led to the change of heart as both global superpowers adhered to Indonesia's request of reclaiming West Papua as part of their republic (Scott and Tebay 2005). As a result, West Papua's independence, which was intended to be facilitated by the Netherlands, was unilaterally halted (Drooglever 2009). The New York Agreement of 1962 was signed with the stipulation that West Papua would be governed temporarily by the United Nations Temporary Executive Authority (UNTEA) before being transferred to Indonesia in 1963 (May 1978). The agreement included a provision that allowed West Papua to have a referendum on whether to become an independent state or remain part of Indonesia. In 1969, a referendum known as the Act of Free Choice was held and it was during this time that the second point of contention of the conflict arose, because only about 1025 West Papuans out of the total population of 800,000 voted in the referendum and all of them voted unanimously in favour of remaining with Indonesia (Saltford 2003). The Indonesian authorities were said to be acutely aware that the majority of West Papuans opposed integration with Indonesia because they desired an independent state, and this was also acknowledged by Adam Malik, Indonesia's Foreign Minister at the time, who was unsure whether the West Papuans would vote to remain in Indonesia (Chauvel and Bhakti 2004). As a result, the referendum process needed to be meticulously planned such that the outcome favoured Indonesia (McGibbon 2006). The 1025 West Papuans who voted in the referendum were allegedly hand-picked to represent the Indonesian interests' and were also subjected to intense pressure from the Indonesian authorities to support integration (Hernawan 2018). Despite the fact that the referendum was held under UN supervision, with Ortiz Sanz leading the delegation, he had apparently expressed grave reservations in his final report to the UN about the ways in which the referendum was implemented (May 1978; Saltford 2003). The absence of indigenous perspectives in the New York Agreement as well as the implementation of the Act of Free Choice without taking the majority voice of West Papuans into consideration, eventually led to initial conflict in the province of West Papua which then snowballed into the current discourse of injustice as well as contemporary arguments for self-determination by many activists, pro-independence groups, and many West Papuan citizens in general due the policies of the different administrations in Indonesia which will be explored further in the following section.

4 The West Papuan Conflict Under the Different Administrations in Indonesia

The situation in West Papua was at its worst during the Suharto administration as the province was declared a military operation zone after the completion of the Act of Free Choice in 1969 (Aspinall and Berger 2001). The Suharto administration deployed aggressive military operations such as execute-style killings and aerial bombings to control the resistance of various groups namely the Free Papua Movement (OPM) that were engaged in independence movements (Osborne 1985). The military operations had also forced many local villagers to evacuate and seek refuge in neighbouring Papua New Guinea (May 1986). Continuous conflict escalation in the province were also prevalent due the destructions of spiritual landmarks of the indigenous society and the seizure of lands from the local community by the Indonesian security forces (Wilson 2001). Customary land ownership is sacred and highly valued by the West Papuan society as a whole, therefore these irresponsible acts carried out by the Indonesian authorities were indeed sensitive, resulting in an anti-Indonesian sentiment within the West Papuan community (Kadir and Ya'kub 2015). Additionally, Suharto's transmigration policy also enraged the local West Papuan community which undoubtedly contributed to the increased tensions in the province, as former residents of Java, Bali, and other provinces who eventually settled in West Papua were said to have received the majority of the economic and job opportunities (Suriadireja 1987). The majority of job opportunities, particularly higher-level positions especially in mining companies such as Freeport McMoran, a US-based mining company operating in West Papua were offered to new residents and non-West Papuans rather than local West Papuans themselves which infuriated them further (Wilson 2001). The argument for this was that local West Papuans lacked the necessary skills for many of the top-level job opportunities with the new residents seen as being more qualified due to their diverse skill sets (Chauvel and Bhakti 2004). Although local West Papuans might have lacked the requisite abilities for some of the jobs, instead of discriminating them, the sensible course of action would have been to organise training sessions so that they learn and develop the necessary skills for the various job opportunities.

However, there were some improvements to the West Papuan conflict post-Suharto with the opening up of political space during the Bacharuddin Jusuf (BJ) Habibie and Abdurrahman Wahid (Gus Dur) administrations, with both leaders acknowledging that the West Papuans were indeed oppressed during the Suharto administration and that Indonesia should bear some responsibility for the current situation in West Papua (Buchanan and Copper 2011). During his brief presidency, BJ Habibie convened a national forum with 100 West Papuan NGOs, church leaders, and intellectual leaders to explore the future of West Papuans as part of his initiative to open up political space for West Papuans. Gus Dur supplemented Habibie's initiative during his administration by facilitating the West Papuan movement's organizational development through the establishment of the Papuan Presidum Council (PDP), which went on to organize two Papuan congresses, one of which was financially supported by the Gus Dur administration. Both congresses discussed indigenous rights, historical legitimacy, and requested the Indonesian government to engage in more dialogues with West Papuan leaders (Kirksey 2012). Nonetheless, Gus Dur's liberal approach of providing more democratic space for the West Papuan movement was unfortunately exploited, with one of the congresses serving as a platform to demand for West Papua's independence (King 2004). Gus Dur was lambasted by several Indonesian politicians including the military forces for his policy in granting too much freedom to the West Papuan movement. Due to the internal and external pressure on his administration, Gus Dur was forced to declare both the Papuan Congresses illegitimate as they were perceived to have threatened Indonesian sovereignty. In addition, the Gus Dur administration also launched a massive military and police operation in West Papua as a precaution to prevent the independence movement from expanding further.

The Megawati Sukarnoputri administration's policy on West Papua began well with the signing of the law on Special Autonomy for West Papua, which theoretically granted the West Papuan provincial government power in all areas except external defence, foreign affairs, judiciary, and monetary policy (Tebay 2005). However, the administration seemed reluctant and were not fully committed in enforcing the law due to fears of West Papuan disintegration and it was also obvious that Megawati had a very similar stance on West Papua to her father, Suharto; as a result, Megawati and her administration's approach towards West Papua remained dormant (McGibbon 2006). The Susilo Bambang Yudhoyono (SBY) administration assumed the duties of the preceding administration by establishing the Papuan Peoples' Assembly (MRP), an official government entity tasked with safeguarding and advancing Papuan values as well as interests guaranteed under the law on Special Autonomy (Singh 2017). The SBY administration was also advocated for consistent dialogue on the issue of West Papua through the Unit for the Acceleration of Development in Papua and West Papua, but despite SBY's best efforts throughout his term in office, he was unable to fully implement the law on Special Autonomy for West Papua (Buchanan and Copper 2011). Despite the fact that the Megawati and SBY administrations expressed concern and some interests in resolving issues as well as conflicts in the province of West Papua, there were also enhanced policies on patrolling local West Papuans either through partitioning or zoning in order to break their resistance by not allowing them to organise as a single unit in addition to enhancing militarization with an increase of approximately 12,000 troops being stationed in West Papua to restrict the freedom of movements among local West Papuans (Hedman 2007). The patrolling and militarisation of the province contributed to the fear among local West Papuans as their way of life were closely guarded by security personnel, which was certainly concerning given the security forces' reputation for carrying out brutal acts against civilians, resulting in internal displacements, that have in fact led to conflicts between local West Papuans and the Indonesian authorities (Wing and King 2005).

The Joko Widodo (Jokowi) administration was marked by promising developments on the issue of West Papua as he called upon responsible parties to end the West Papuan conflict through constructive dialogue in addition to releasing at least 32 West Papuan political prisoners and the commitment to lifting restrictions in allowing foreign journalists to have access into West Papua (Chauvel 2019). Jokowi also urged all the ministries responsible for infrastructure development in his administration to accelerate the constructions of facilities such as highways and airports in West Papua (Sundaryani 2017). The Jokowi administration also offered West Papua the opportunity to host the Indonesian National Games in October 2021, demonstrating the administration's commitment to developing the province and assuring the public that the province was indeed a safe place contrary to the general public perception (Mambor 2021). Furthermore, it is also worth noting that Jokowi visited the province of West Papua fourteen times, more than any previous Indonesian presidents, demonstrating his commitment to the province although more improvements are needed in regards to policies on West Papua (West Papua Daily 2022). However, despite the many positives, Jokowi's policy towards West Papua has also raised some eyebrows, especially in his handling of the large anti-racism protest in the West Papuan cities of Jayapura, Timika, and Fakfak in 2019. The protest occurred in response to footage displaying Indonesian security personnel referring to West Papuan students in Surabaya as *monyet* (monkeys), a racist term long used by Indonesians to ridicule the physical appearance of West Papuans (Okthariza 2019; Shelton and Wibawa 2019). Instead of taking action against the security personnel responsible and reassuring local West Papuans, the Jokowi administration deployed 6,000 troops into West Papua to brutally supress the protest (Arndt 2019), resulting in the imprisonment of 36 protestors for treason (Firdaus 2020). The appointment of former General Prabowo Subianto, as Minister of Defense in October 2019 by the Jokowi administration was also another questionable decision as it proved to be very unpopular among many West Papuans due to the fact that Prabowo had a history of being implicated in multiple human rights violations, particularly in the massacre of West Papuans civilians in 1996 (Tasevski 2022) and there was genuine fear that similar actions might be repeated especially when Prabowo has control over the millitary forces as minister. While the Suharto administration was the most difficult for West Papuans, conditions improved marginally under subsequent Indonesia administrations; however, numerous concerns remain in West Papua with issues such as racism, human rights development, and job opportunities for locals that needs to be continuously addressed in the province. The next section explores the formation and the evolution of the MSG as a sub-regional organisation and their involvement in the West Papuan conflict.

5 The Formation and the Evolution of the Melanesian Spearhead Group (MSG)

The MSG is one of the three sub-regional organisations in the Pacific Islands and its members include Melanesian countries from the region. Some of the MSG member states such as Papua New Guinea, Vanuatu and the Solomon Islands are also members of the PIF which sometimes explains the convergence between both organisations. As a sub-regional organisation, the MSG has definitely been active especially in addressing various issues within the Melanesia region. The impetus for the establisment of MSG was solely based on the newly formed Melanesian states' pride in identity, ethnic and cultural unity (Cain 2014). The MSG emerged as a result of an informal meeting in 1986 with the leaders of Vanuatu, Papua New Guinea, Solomon Islands and the representative of the Front de Libération Nationale Kanak et Socialiste (FLNKS), a political party from the French Territory of New Caledonia (MSG 2021). The MSG was then established as a loose coalition of Melanesian states in 1988 with the adoption of the Six Agreed Principles of Cooperation among the independent states in Melanesia. The initial signatories were Vanuatu, Papua New Guinea and Solomon Islands as FLNKS joined in 1989 with Fiji being the final member to join the MSG in 1998 (MSG 1988). The MSG finally became a legal entity under international law in 2007 when all of its member states signed an agreement formally establishing itself as the Melanesian Spearhead Group (MSG), validating its status as an official sub-regional organisation alongside other regional organisations (MSG 2007). Following their formal recognition, the MSG established their secretariat in Vanuatu which eventually became MSG's official headquarters in 2008.

The MSG thrived as sub-regional organisation during its initial establishment in 1988 although it was not recognised yet officially under international law as it was considered an informal entity. Nonetheless despite its informality, the MSG was active in promoting trade and regional integration through the signing of the MSG Trade Agreement (MSG-TA) between the Solomon Islands, Vanuatu, and Papua New Guinea in 1993 which initially covered products and commodities such as tea, canned tuna, and beef (Marawa 2015). With Fiji's inclusion and ratification of the MSG-TA, the agreement was expanded in 1997 to include trade of more than 180 goods and commodities by which during this time, the MSG was looked upon as a formidable entity in the Pacific (Jayaraman 2012). The MSG-TA was indeed a ground-breaking agreement for such a small organisation because there were not many regional organisations let alone informal sub-regional organisations that had a functioning trade agreement. The MSG-TA was revised and signed in 2005 (MSG-TA 2) as it was strengthened and improved to eliminate import tariffs on all products exported with the exception of items excluded by member states for reasons due to health and environmental protection. With the MSG's legality becoming apparent in 2007 and the watershed moment of establishing its own headquarters in 2008, the MSG elevated their cooperation among its member states to a whole new level by revising the MSG-TA 2 in beginning negotiations towards the creation of the Melanesian Free Trade Agreement (MFTA or MSG-TA 3), and it was expected that the agreement would come into force in 2017 (Radio New Zealand 2016). However, for the moment only two member states have signed the MFTA while other member states are in the process of ratifying the agreement although negotiations over the MFTA has been finalised and once the MFTA is ratified by all four MSG member states, it will officially replace the MSG-TA 2. The MFTA is an enlargement of the previous trade agreements (MSG-TA and MSG-TA 2) that primarily covered trade in products, as the new agreement would be augmenting trade in services, cross-border investment, and semi-skilled labour mobility which would indeed further enhance MSG's economic integration (MSG 2018).

In addition to its solid economic cooperation as a sub-regional organisation, the MSG has also demonstrated progress in the area of political cooperation. Since its formation, the MSG has backed the FLNKS in their campaign for New Caledonia's independence from France. New Caledonia is a French-controlled territory that is populated with the Kanak indigenous community as well as Europeans that were born on the island. The Noumea Accords, signed in 1998, offered New Caledonians a referendum on independence from France, among other things (Maclellan 1999). The referendum on independence was carried out three times, and each time majority of New Caledonians overwhelmingly rejected independence and opted to remain as a French territory, despite constant pressures from pro-independence forces (Al-Jazeera 2021). In showing their continuous support and commitment to the FLNKS, the MSG frequently expressed concern about the inactive progress of the Noumea Accords (May 2011) in addition to establishing a FLNKS unit within the MSG Secretariat to assist with the implementation of the provisions within the Noumea Accords that have yet to materialised (Cain 2014). Beyond the FLNKS, the MSG has also increased its visibility in the international community by establishing a peace-keeping section within the sub-regional organization in deploying manpower for UN peacekeeping missions worldwide. Thus, after decades of being viewed primarily as a forum on economic and trade partnerships, the MSG has indeed developed as a regional organization of substance with its focus on sub-regional political cooperation as well as its wider influence within the international community.

The MSG has also undoubtedly enhanced its reputation as a sub-regional organisation by showcasing its role in the area of conflict management with the West Papuan conflict being one of MSG's main focus in the region. Despite the fact that the conflict has often been labelled as an internal conflict within Southeast Asia, the Papuan region as a whole is considered part of Melanesia and the Melanesian solidarity as well as identity has been an important motivation for the MSG in pursuing their active role in the West Papuan conflict. In the next section, the author goes into the details of MSG's role in the West Papuan conflict by analysing the many contributions and efforts carried out by the sub-regional organisation through preventive diplomacy, highlighting some of the preventive diplomacy strategies utilised in addition to arguing that the MSG has indeed been able to minimise the West Papuan conflict despite the numerous recurring challenges surrounding the province as a whole.

6 The MSG's Contributions to the West Papuan Conflict Through Preventive Diplomacy

In this section, the author argues that the MSG played an influential role as a sub-regional organization in managing the West Papuan conflict. The MSG's management of the conflict is analysed through the preventive diplomacy framework by integrating MSG's various contributions and efforts along with the different preventive diplomacy strategies implemented that led to some form of minimisation in the West Papuan conflict. One of the MSG's first contribution in addressing the West Papuan conflict was to ensure that the West Papuan issue is constantly and consistently discussed as one of the fixed agenda item at MSG summits and meetings. The issue of West Papua began to be actively addressed at the 19th MSG Leaders' Summit in 2013, during which the membership application of West Papua to the MSG was at the forefront of the summit with discussions taking place after the proposal was made through the West Papuan National Council for Liberation (WPNCL). During the same summit, the Vice Chairman of the WPNCL, Dr Otto Ondawame was given the opportunity and platform to represent West Papua in delivering a speech to the members of the summit in which he addressed the concerns of the West Papuan situation (Ondawame 2013). This was indeed a huge endorsement by the MSG delegation as it provided West Papua with some form of legitimation to address the struggle of the West Papuan community. The MSG also expressed their concern over the human rights violations and killings in the province of West Papua which demonstrated the MSG's sincerity and commitment in addressing the plight of the West Papuans (MacLeod 2013). TheWest Papuan issue became an important meeting agenda once again during the 20th MSG Leaders' Summit held in 2015 in which the MSG leaders met to discuss about the possibility of granting West Papua membership at the MSG. The discussion about West Papua's membership was apparently labelled as the summit's most high-profile issue discussed among MSG leaders, demonstrating West Papua's importance to the MSG and their seriousness in ensuring that West Papua-related issues are prioritized at high-level MSG meetings and summits (Mote 2015).

Through MSG's contribution in ensuring that the West Papuan issue is consistently addressed at MSG summits and meetings, I observe MSG's role in using the council of the wise as a preventive diplomacy strategy in managing the West Papuan conflict. While there is an understanding that the council of the wise serves as a stand-in for a cultural council in mediating and managing conflicts, I consider the MSG to be a cultural council in this instance because of its Melanesian identity, which has provided West Papua with a platform to showcase their struggles with their fellow Melanesian counterparts while also having consistent discussions about the ongoing conflict and issues that needs to be addressed in West Papua. The MSG managed the conflict by ensuring that the issue of West Papua was a priority item on the agenda at MSG meetings and summits. In the view of the concerns expressed by their West Papuan counterparts, the MSG could then attempt to provide sufficient recommendation in terms of the next course of action in addition to sustaining the momentum of discussions on the difficulities faced by West Papuans at subsequent MSG meetings which could lead to some form of agreement in order to avoid escalating the conflict. The platforming of West Papua by the MSG is unquestionably a step forward in minimising tensions in the province and between the conflicting parties. From the West Papuan perspective, the MSG's use of the council of the wise as a preventive diplomacy strategy has somewhat played a significant role in showing down the conflict in West Papua when the issue of West Papua became an important agenda item at the sub-regional level especially with Indonesia's presence at these meetings however precautions should be exercised on some sensitive topics, such as the issue of self-determination which may trigger Indonesia's refusal to dialogue or have further discussions with the representatives from West Papua.

The MSG's second contribution in addressing the West Papuan conflict was its acceptance of the United Liberation Movement for West Papua (ULMWP) as an observer to the MSG in 2015 effectively granting them somewhat of a sub-regional membership. The ULMWP was formerly known as the WPNCL, which first submitted its membership request to the MSG through the Prime Minister of Fiji in April 2013 (MSG 2013). The membership was rejected because the MSG questioned the representation of the WPNCL (MacLeod et al. 2016) as many of the organisations under the WPNCL were not locally based and the MSG was concerned that the voice of the WPNCL did not represent the grassroots in West Papua (Andrews 2015). As a result of the Special MSG Summit in 2014 that was extensively discussing the West Papuan membership, the MSG member states collectively decided that representatives from West Papua would need to establish a united umbrella organisation and submit a new membership application to the MSG (MacLeod et al. 2016). This new organisation needed to include West Papuans of various backgrounds and ideologies including those that were not in favour of West Papuan independence. In responding to MSG's request for the formation of a new collective, a unification conference was organised to foster better communication and stronger cooperation among West Papuan leaders from various backgrounds, which then resulted in the establishment of the ULMWP in December 2014. The unification conference was a collaborative effort between the MSG, the government of Vanuatu which sponsored the conference and the Pacific Council Churches which brought together many West Papuan leaders (Macleod 2015). The newly established ULMWP applied for MSG membership, which was subsequently evaluated during the 20th MSG Leaders' Summit in June 2015, with all the MSG member states unanimously agreeing to grant the ULMWP observer status at the MSG (Fox 2015). Despite the fact that the ULMWP did not obtain full membership, the Secretary General of ULMWP, Octovanius Mote voiced his satisfaction with the outcome as he considered the observer status as a first step towards full membership in the future (Gilio-Whitaker 2015). MSG leaders from Vanuatu, Solomon Islands, and the FLNKS even reaffirmed their commitment in ensuring that West Papua through the ULMWP is granted full membership at the MSG in the future (Yapumi 2016). The observer status granted to West Papua under the ULMWP at the MSG was indeed a significant milestone because it was considered to be West Papua's first ever diplomatic recognition (Free West Papua 2014) which then enables issues pertaining to West Papua to be highlighted more frequently. The observer status of the ULMWP also allows it to lobby Indonesia to take the appropriate steps to address the human rights violations in West Papua.

By granting the rebranded ULMWP observer status at the MSG, I recognised MSG's role in employing norms and rules as a preventive diplomacy strategy in managing the West Papuan conflict. As mentioned previously in the earlier section of this chapter, the purpose of this strategy is to encourage compliance with certain standards and rules in order to acquire regional or international acceptability. In this context, the West Papuan leaders and representatives were required to conform with the MSG in order to obtain regional authorisation for their participation in the MSG-led sub-regional organization. The WPNCL's initial membership application was declined because the organisation was perceived to be deficient in diversity in terms of having West Papuan representations from different backgrounds and political ideology, notably those opposed to the West Papuan self-determination. The MSG acknowledged this limitation and proposed that the West Papuan leaders expand their composition so that their overall grouping did not appear biased by only representing a particular segment of the West Papuan society. In full compliance to the MSG, the West Papuan leaders convened a conference and founded the ULMWP, which went on to obtain observer status at the MSG. It was essential that the MSG ensured that even those opposed to West Papuan independence were represented in the ULMWP as part of a new umbrella organisation that did not appear to be discriminatory at the MSG level. Furthermore, it was also vital that the West Papuan representatives at the MSG were acknowledged by Indonesia allowing for improved communication between both parties to address their differences in relation to the conflict in West Papua. The unfair balance of the West Papuan representatives at the MSG would have led to disagreements both within the West Papuan society as well as with Indonesia because discussions at the MSG would have mostly been about self-determination rather than other pressing issues thus the MSG's move to avoid this particular episode from the very beginning should be commended and it also proved to have momentarily minimised the conflict in West Papua.

The MSG's third contribution in addressing the West Papuan conflict was by strengthening its relations as a sub-regional organization with Indonesia. It was essential for the MSG to maintain as well as enhance its relationship with Indonesia, particularly in relation to the issue of West Papua because disregarding Indonesia would eventually put an end to conversations surrounding the conflict in West Papua, denying the MSG access to information especially in regards to the challenges and struggles facing many West Papuans. One of the initial steps taken by the MSG to improve relations with Indonesia was when it agreed collectively as an organisation to grant Indonesia observer status at the MSG in March 2011 (Webb-Gannon and Elmslie 2014). The MSG-Indonesia relations was boosted further when Indonesia extended an invitation to the MSG leaders to visit West Papua and inspect the human rights situation on the ground, which resulted in the MSG's Foreign Ministers Delegation visiting West Papua in January 2014 (Poling 2013; MacLeod et al. 2016). Despite the fact that the delegation did not engage with any civil society organizations or West Papuan leaders that were lobbying for self-determination, the overall visit was positive as the delegation managed to have engagements with police officials in Jayapura in addition to also witnessing a protest by some West Papuan activists (Blades 2014). The MSG-Indonesia relations deepened even further when Indonesia was granted associate membership at the MSG during the 20th MSG Leaders' Summit in 2015 (Fox 2015). As an associate member, Indonesia will be represented at MSG meetings by governors from the five Melanesian provinces of Papua, West Papua, Nusa Tenggara, North Maluku, and South Maluku (Carter and Firth 2015). In this context, it is important to note that Papua and West Papua began to be administered by Indonesia as two separate provinces from 2003 onwards as the whole province prior to the division was known as West Papua. The MSG's decision to upgrade Indonesia's membership from observer to associate member was both practical and sensible, since the five provinces mentioned were unquestionably home to a sizable Melanesian population.

The MSG's effort of enhancing its relations with Indonesia, very much showcases its role in implementing two preventive diplomacy strategies of formal mediation and fact finding in address the situation in West Papua. In terms of formal mediation, the MSG leverages on the organisations Melanesian identity in trying to influence as well as acquire the trust of the conflicting parties of both West Papua and Indonesia. However in this case, the MSG has placed a special emphasis on acquiring Indonesia's trust because, it understands that without a strong relationship with Indonesia it would be nearly impossible to address the situation in West Papua considering the fact that the province is legally part of Indonesia. The MSG's formal mediation process was carried out by offering MSG membership to both West Papua (observer) represented by the ULMWP, and Indonesia (associate member) represented by the five provinces with Melanesian populations, as a way of brokering more direct communications between the conflicting parties that are facilitated by the MSG during meetings and summits. Although some might argue that Indonesia has greater influence as an associate member, the reality is that both the ULMWP and Indonesia are equals within the organisation because neither conflicting parties were granted full membership therefore limiting their ability to dominate proceedings during MSG meetings and summits. Furthermore, the MSG members are the ones that usually lead the proceedings and are responsible for setting the agenda during MSG meetings and summits in addition to facilitating discussions between the ULMWP and Indonesia especially on issues relating to West Papua. The MSG indeed has a upper hand in the mediation process involving both parties in making sure that the conflict between both parties does not reach a boiling point. Under the same backdrop of improving relations with Indonesia, the MSG also adopted fact finding as a preventive diplomacy strategy. This strategy was only conceivable after Indonesia invited MSG leaders to visit West Papua. In this context the convergence between both strategies of formal mediation and fact finding was very much crucial because the MSG's visit to West Papua would not have materialised without MSG's initial role in formal mediation which then encouraged Indonesia to invite the MSG leaders to visit the province. The MSG Foreign Ministers definitely took the opportunity by accepting Indonesia's invitation as they visited West Papua in 2014 and it was indeed a significant visit because it became sort of a fact-finding mission because the MSG was able to gain an understanding of the situation in West Papua. Although the delegates missed out meeting with civil society representatives and pro-independence West Papuan leaders, the visit was important for gathering valuable information and data about the situation on the ground in West Papua which then influenced the MSG's decision making process in addressing the conflict more efficiently. Overall, both the formal mediation process and fact finding missions facilitated continuous dialogues between the ULMWP and Indonesia which contributed to the minimisation of the conflict in West Papua.

The MSG's fourth contribution in addressing the West Papuan conflict was to raise more awareness about the West Papuan struggle through international and regional platforms such as the United Nations (UN), the Pacific Islands Forum (PIF) and the Africa Caribbean Pacific (ACP) Group. As MSG member states,, Vanuatu and the Solomon Islands have consistently raised the issue of human rights violations committed by Indonesian authorities in West Papua on numerous occasions in various international and regional platforms. Former Vanuatu Prime Minister Moana Carcasses who spoke at the UN General Assembly in September 2013, advocated for the appointment of a UN special representative to examine and investigate human rights violations in West Papua (Wirz 2013). He also expressed his concern about West Papua again to the UN Human Rights Council in Geneva, in March 2014, underlining the ongoing human rights misconducts committed by the Indonesian intelligence unit in the form of torture, rape, and military raids (Free West Papua Campaign 2014; Lawson 2016). The Chargé d'affaires at the Permanent Mission of Solomon Islands to the UN in Geneva, Barrett Salato in March 2016 also addressed the worsening human rights situation in West Papua by emphasizing the importance of raising up the issue of West Papua to the UN for the purpose of awareness because many in the international community apparently do not possess enough information about the situation in the province (Solomon Times 2016). The consistency by Vanuatu and the Solomon Islands was demonstrated further when they along and five other Pacific nations consisting of Tonga, Tuvalu, Palau, Nauru, and the Marshall Islands in March 2017, urged the UN to investigate alleged human rights violations in West Papua (Fox 2017). This was indeed an important show of solidarity by the Pacific nations alongside the two MSG member states as they genuinely recognised the seriousness of the situation in West Papua. The issue of West Papua was also raised by the PIF which includes four of the five MSG member states, when the regional organization in August 2019, issued a joint communique strongly encouraging Indonesia to finalize a specific timing for the visit by the Office of the High Commissioner of Human Rights to investigate reported human rights violations in West Papua, that apparently included acts of torture, extrajudicial killings, and systemic police and military violence (Lyons and Doherty 2019). Following that, the 79-member states of the Africa Caribbean Pacific (ACP) Group which includes four MSG member states in December 2019, passed a resolution calling for the urgent attention of the human rights situation in West Papua (Radio New Zealand 2019). Indonesia's Universal Periodic Review (UPR) at the UN in November 2022, was one of the most recent occasions in which Vanuatu, an active MSG member state was involved in highlighting the West Papuan issue by calling out Indonesia to accept the visit of the Office of the High Commissoner of Human Rights immediately to the provinces of Papua and West Papua without further delays (Ibnu Aqil 2022). The UPR is an essential instrument that allows UN agencies, member states and civil society organisations to evaluate a state's human right progress by ensuring accountability. The UPR also demonstrated that in addition to Vanuatu, even other influential states such as Australia and New Zealand as members of the PIF have concerns over the human rights situation in West Papua in addition to the United States that had to also advise Indonesia to conduct investigations into all allegations of extrajudicial killings in West Papua and to take appropriate actions against the perpetrators (Ibnu Aqil 2022).

The promotion of more awareness in addition to calling out some of the violations by the Indonesian authorities in West Papua very much demonstrates, MSG's role in utilising legal frameworks and agreements as a preventive diplomacy strategy in ensuring that the issue of West Papua becomes an interest at the UN and among the different regional platforms particularly those that are influenced by the MSG member states. While some might suggest that calling out Indonesia on the international stage would trigger an escalation to the West Papuan conflict, I would argue on the contrary because there is a significant reputational risk if Indonesia does not respond to the various concerns that has been raised on a consistent basis, particularly at the international level. As MSG member states, both Vanuatu and the Solomon Islands were able to pressure Indonesia to act in light of the revelations about human rights violations through legal frameworks available at the UN, with Vanuatu in particular opting to address the West Papuan issue through the UN Human Rights Council in 2014, where both Vanuatu and Indonesia were members. The council certainly provided an appropriate platform for Vanuatu, through its former Prime Minister Moana Carcasses in 2014, to address the human rights violations within the legal framework of a UN entity which Indonesia was already a part of in accordance with the legal agreements guiding the council in which members were somewhat obligated to observe in addition to responding to the concerns brought up by other members as a way of managing their reputation as a state within a particular organisation. Ultimately, the urgency by Vanuatu in addition to the intervention by other members within the UN Human Rights Council should have definitely influenced Indonesia to take the necessary measures in managing the situation in West Papua and preventing it from escalating further. The use of legal frameworks was also demonstrated when Vanuatu and the Solomon Islands along with the five Pacific nations urged the UN to investigate human rights violations in West Papua, to which Indonesia responded by agreeing in principle to allow the Office of the High Commissioner for Human Rights to visit West Papua (Davidson 2019). This was followed by Vanuatu demanding that Indonesia approve the visit without any further delay during Indonesia's UPR at the UN in November 2022. However it remains to be seen whether Indonesia reacts to the demand but nonetheless a significant aspect of the UPR is that the issue of West Papua was also raised by Australia, New Zealand and the United States. This highlights the magnitude of the human rights situation in West Papua, as it has garnered the attention of major states thereby emphasizing the need for Indonesia to take effective measures in managing the conflict. The MSG through its members states, has also ensured that Indonesia adheres to its commitment on West Papua by posing questions through various legal agreements too such as the issuance of joint communiques and resolutions issued by the PIF and ACP respectively, which I believe sent a strong message and contributed to the minimization of the conflict in West Papua. Although Indonesia's response to the criticisms and recommendations by the MSG member states and its partners, are sometimes viewed as just statements without genuine intention, the fact that Indonesia responds to the various criticism itself demonstrated its effort in avoiding the credibility question.

This suggests that Indonesia is taking steps to manage the conflict in West Papua especially when more actors are becoming involved in addressing the human rights concerns in the province. Therefore, MSG's initiative in invoking legal frameworks and agreements as a preventive diplomacy strategy to raise awareness about the conflict in West Papua have resulted in some compliance by Indonesia at least for a period of time due to the reputational factor which has contributed to the de-escalation of the West Papuan conflict.

7 Conclusion

Despite being a sub-regional organisation, the MSG'S role in the West Papuan conflict has indeed been instrumental. Through the preventive diplomacy framework, the author was able to provide a systematic analysis of the MSG's significant role in the West Papuan conflict through the numerous contributions and efforts by highlighting the utilisation of preventive diplomacy strategies that seemed to have been influential in managing the conflict in West Papua. The MSG initiated four key contributions that aligned with the implementation of five strategies under the preventive diplomacy framework. These contributions and strategies brokered positive engagements and dialogues between West Papua and Indonesia at the sub-regional level which enabled conflict minimisation in the province. One of MSG's first role in contributing to the West Papuan conflict involves being a cultural council by emphasising the importance of the Melanesian identity in ensuring that the West Papuan issue is discussed consistently and constantly in addition to being a fixed agenda item at the regional level during MSG summits and meetings. Next, the MSG also demonstrated its role as the facilitator of norm and rules by granting the ULMWP membership as an observer at the sub-regional level. This was an important step as it diversified the West Papuan representation at the MSG and prevented the exclusion of those who were not advocating for West Papuan independence. The MSG's insistence on compliance to its established regulations regarding membership diversification of the West Papuan representatives was instrumental in avoiding disagreements with Indonesia especially on the issue of West Papuan independence which contributed to the minimisation of the conflict between West Papua and Indonesia. The MSG also demonstrated its role in formal mediation by enhancing its relations with Indonesia. The MSG once again leveraged on the organisations Melanesian identity in trying to influence and establish trust among the contending parties in both West Papua and Indonesia with a particular emphasis on Indonesia, recognising the fact that discussions on the issue of West Papua would be impossible without the involvement of Indonesia. The successful mediation with Indonesia eventually culminated in the MSG demonstrating its role in fact finding, with Indonesia extending an invitation to the MSG leaders to visit West Papua. The MSG Foreign Ministers visit to West Papua became a fact-finding mission, as it enabled the MSG to comprehend the realities on the ground which ensured that conversations about the West Papuan conflict continued amongst the conflicting parties, directly leading to the minimisation of the conflict. The final assessment of the MSG's contribution to the West Papuan conflict is its role in the employment of legal frameworks and agreements that resulted in the global and regional awareness of the West Papuan conflict. While it appeared that Indonesia would react negatively to the MSG citing international legal provisions on the West Papuan issue, I very much maintained a contrary view due to the reputation risk observed, particularly when Indonesia responded to the calls by MSG member states in addressing some of the human rights issues and concerns. Overall, the MSG's influential role in the West Papuan conflict has indeed been proven to be crucial and it has added a new dimension in the research area of conflict management and regional organisations particularly in examining the function of sub-regional organizations in the Melanesian region.

References

Abuza Z (2019) [ANALYSIS] Philippine defense spending in 2018: what data tells us. Rappler. https://www.rappler.com/voices/thought-leaders/235184-analysis-philippine-defense-spending-2018-sipri-data-tells-us/

Al-Jazeera (2021) New Caledonia rejects independence from France in third vote. Al-Jazeera. https://www.aljazeera.com/news/2021/12/12/new-caledonia-rejects-independence-from-france-in-third-vote

Andrews S (2015) West Papua: Melanesian Spearhead Group has a tough decision to make. The Interpreter. https://www.lowyinstitute.org/the-interpreter/west-papua-melanesian-spearhead-group-has-tough-decision-make

Arndt P (2019) We are not monkeys! ARENA. https://arena.org.au/we-are-not-monkeys-by-peter-arndt/

ASEAN Today (2021) Thailand's deep south conflict poses a problem for pro-democracy reform | ASEAN Today. https://www.aseantoday.com/2021/03/thailands-deep-south-conflict-poses-a-problem-for-pro-democracy-reform/

Aspinall E, Berger MT (2001) The break-up of Indonesia? Nationalisms after decolonisation and the limits of the nation-state in post-cold war Southeast Asia. Third World Q 22(6):1003–1024

Babbitt EF (2012) Preventive diplomacy by intergovernmental organizations: learning from practice. Int Negot 17(3):349–388

Blades J (2014) Melanesia's test: the political quandary of West Papua. Pac Journal Rev 20(2):23

Borda S (2011) Optimizing preventive-diplomacy tools: a Latin American Perspective. In: Preventive diplomacy: regions in focus. International Peace Institute, pp 15–27. https://www.ipinst.org/2012/01/preventive-diplomacy-regions-in-focus

Boutros-Ghali B (1992) An agenda for peace. Int Relat 11(3):201–218

Buchanan C, Copper A (eds) (2011) Conflict Management in Indonesia—An Analysis of the Conflicts in Maluku, Papua and Poso. Centre for Humanitarian Dialogue

Cain TN (2014) The Melanesian Spearhead Group: what is it, and what does it do? | Lowy Institute. The Interpreter. https://www.lowyinstitute.org/the-interpreter/melanesian-spearhead-group-what-it-what-does-it-do

Carter G, Firth S (2015) The mood in Melanesia after the Regional Assistance Mission to Solomon Islands. Asia Pacific Policy Stud 3(1):16–25

Chauvel R (2019) Papua under the Joko Widodo Presidency. In: Continuity and change after Indonesia's reforms contributions to an ongoing assessment. ISEAS–Yusof Ishak Institute

Chauvel R, Bhakti IN (2004) The Papua conflict: Jakarta's perception and policies. East-West Center

Cordovez D (1987) Strengthening United Nations diplomacy for peace: the role of the secretary general. In: Renniger J (ed) The United Nations and the maintenance of international peace and security. Martinus Nijhoff Publishers
Cohen R, Hoffman JS, Knox LM (2010) Beyond suppression: global perspectives on youth violence. Praeger
Cristescu R, Nicolescou A, Wandi A (2012) ASEAN and peace mediation: progress, challenges, and cooperation. Crisis Management Initiative (CMI). https://reliefweb.int/report/world/asean-and-peace-mediation-progress-challenges-and-cooperation
Davidson H (2019) Indonesia to let UN workers into West Papua as violence continues. The Guardian. https://www.theguardian.com/world/2019/jan/30/indonesia-to-let-un-workers-into-west-papua-as-violence-continues
Della-Giacoma J (2011) Preventive diplomacy in Southeast Asia: redefining the ASEAN way. In: Preventive diplomacy: regions in focus. International Peace Institute, pp 28–34. https://www.ipinst.org/2012/01/preventive-diplomacy-regions-in-focus
Drooglever PJ (2009) An act of free choice: decolonization and the right to self-determination in West Papua. Oneworld
Emmers R, Tan SS (2009) The ASEAN Regional forum and preventive diplomacy: a failure in practice. RSIS Working Paper, 189:1–21
Firdaus F (2020) Seven Papuan activists convicted of treason after anti-racism protests. Al-Jazeera. https://www.theguardian.com/global-development/2020/jun/17/seven-papuan-activists-convicted-of-treason-after-anti-racism-protests
Fox L (2015) West Papuans given Melanesia bloc access, but not full membership. ABC News. https://www.abc.net.au/news/2015-06-27/west-papuans-given-melanesia-bloc-access/6577722
Fox L (2017) Pacific nations call for UN investigations into West Papua. ABC News. https://www.abc.net.au/news/2017-03-02/pacific-nations-call-for-un-investigations-into-west-papua/8320194
Free West Papua Campaign (2014) Vanuatu PM speaks out for West Papua at UN again. Free West Papua Campaign. https://www.freewestpapua.org/2014/03/04/vanuatu-pm-speaks-out-for-west-papua-at-un-again/
Fröhlich M (2018) Dag Hammarskjöld, 1953–1961. In: Fröhlich M, Williams AW (eds) The UN secretary-general and the security council: a dynamic relationship. Oxford University Press
Gilio-Whitaker D (2015) West Papua wins observer status in Melanesian Spearhead Group. Intercontinental Cry. https://intercontinentalcry.org/west-papua-wins-observer-status-in-melanesian-spearhead-group-28313/
Hara F (2011) Preventive Diplomacy in Africa: adapting to new realities. In: Preventive diplomacy: regions in focus. International Peace Institute, pp 4–14. https://www.ipinst.org/2012/01/preventive-diplomacy-regions-in-focus
Hedman E-L (ed) (2007) Dynamics of conflict and displacement in Papua, Indonesia. RSC Working Paper, 42
Hernawan B (2018) Torture and peacebuilding in Indonesia: the case of Papua. Routledge
Ibnu Aqil AM (2022) Indonesia faces the music at UN human rights review. The Jakarta Post. https://www.thejakartapost.com/world/2022/11/10/indonesia-faces-the-music-at-un-human-rights-review.html
Jayaraman TK (2012) How MSG leads the trade way. Fiji Sun. http://tkjayaraman.com/docs/2012/np_articles/How%20MSG%20leads%20the%20trade%20way_FIJI%20SUN_Saturday%20October%2013%202012.pdf
Jenča M (2013) The concept of preventive diplomacy and Its application by the United Nations in Central Asia. Secur Hum Rights 24(2):183–194
Kachuyevski A (2012) The possibilities and limitations of preventive action: the OSCE high commissioner on national minorities in Ukraine. Int Negot 17(3):389–415. https://doi.org/10.1163/15718069-12341237
Kadir M, Ya'kub A (2015) Revisiting self determination conflicts in Indonesia: an international law perspective. Indonesia Law Rev 5(2):123–139

Kim HS (2006) The determinants of internal conflict in the third world. Whitehead J Dipl Int Relat 7(91):91–106. https://heinonline.org/HOL/LandingPage?handle=hein.journals/whith7&div=12&id=&page=

King P (2004) West Papua & Indonesia since Suharto: independence, autonomy or chaos? UNSW Press

Kirksey E (2012) Freedom in entangled worlds. Duke University Press

Knight WA, Yamashita M (1993) The United Nations' contribution to international peace and security. In: Building a new global order: emerging trends in international security. Oxford University Press

Lagerberg K (1979) West Irian and Jakarta imperialism. C. Hurst & Company

Lawson S (2016) West Papua, Indonesia and the Melanesian Spearhead Group: competing logics in regional and international politics. Aust J Int Aff 70(5):506–524

Legge JD (2004) Sukarno: a political biography. Archipelago, Northam

Lijphart A (1966) The Trauma of decolonization: the Dutch and West New Guinea. Yale University Press, New Haven

Lund M (2001) Preventing violent conflicts: a strategy for preventive diplomacy. United States Institute of Peace Press

Lyons K, Doherty B (2019) West Papua: pacific leaders urge UN visit to region's "festering human rights sore." The Guardian. https://www.theguardian.com/world/2019/aug/16/west-papua-pacific-leaders-urge-un-visit-to-regions-festering-human-rights-sore

Maclellan N (1999) The Noumea accord and Decolonisation in New Caledonia. J Pacific Hist 34(3):245–252

MacLeod J (2013) A win for West Papua in Melanesia. New Matilda. https://newmatilda.com/2013/07/01/win-west-papua-melanesia/

Macleod J (2015) Merdeka and the morning star: civil resistance in West Papua. University of Queensland Press

MacLeod J, Moiwend R, Pilbrow J (2016) A historic choice: west Papua, human rights, and pacific diplomacy at the pacific Island Forum and Melanesian Spearhead Group. https://www.ulmwp.org/wp-content/uploads/2016/09/WP_PIF_MSG_Report_Online_RLR-1.pdf

Mambor V (2021) Indonesia's Papua Region Hosts National Games Amid Rise in Insurgency. Benar News. https://www.benarnews.org/english/news/indonesian/indonesia-holds-national-games-in-papua-10012021145325.html

Mancini F (2011) Preventive diplomacy: regions in focus. International Peace Institute

Marawa S (2015) Negotiating the Melanesia free trade area. In: Fry G, Tarte S (eds) The new pacific diplomacy. ANU Press, pp 161–174

May B (1978) The Indonesian tragedy. Graham Brash (Pte) Ltd.

May RJ (1986) Between two nations: The Indonesia—Papua New Guinea border and West Papua Nationalism. Robert Brown and Associates

May R (2011) The Melanesian spearhead group: testing pacific Island solidarity. ASPI Policy Analysis 74:1–8

McGibbon R (2006) Pitfalls of Papua: understanding the conflict and its place in Australia-Indonesia relations. Lowy Institute for International Policy

Moon BK (2011) S/2011/552. Report on "Preventive Diplomacy: Delivering Results". United Nations, pp 1–26

Mote O (2015) 20th MSG leaders summit statement by Octovianus mote secretary general of the ULMWP [In-Person]. 20th MSG Leaders' Summit. https://www.msgsec.info/wp-content/uploads/Statements/2015-Jun-20th-MSG-Leaders-Summit-Statement-by-Octovianus-Mote-Sec.-General-ULMWP.pdf

MSG Secretariat (1988) About MSG—Signing of the agreed principles of cooperation among MSG member states, Port Vila, Republic of Vanuatu. Melanesian Spearhead Group Secretariat. https://msgsec.info/about-msg/

MSG Secretariat (2007) About MSG—Agreement establishing the MSG 2007. Melanesian Spearhead Group Secretariat. https://msgsec.info/about-msg/

MSG Secretariat (2013) MSG Chairman meets with the West Papua National Coalition for Liberation. Melanesian Spearhead Group Secretariat. https://msgsec.info/historical-news/
MSG Secretariat (2018) Melanesian Spearhead Group Annual Report 2017. In: Melanesian Spearhead Group secretariat. Director General Melanesian Spearhead Group (MSG) Secretariat, pp 1–70. https://msgsec.info/annual-reports/
MSG Secretariat (2021) About MSG. Melanesian Spearhead Group Secretariat. https://msgsec.info/about-msg/
Muggah R, White N (2013) Is there a preventive action renaissance? The policy and practice of preventive diplomacy and conflict prevention. Norwegian Peacebuilding Resource Centre (NOREF), pp 1–13. https://reliefweb.int/report/world/there-preventive-action-renaissance-policy-and-practice-preventive-diplomacy-and
Murithi T, Mwaura C (2010) The Panel of the Wise. In: Engel U, Porto JG (eds) Africa's new peace and security architecture: promoting norms, institutionalizing solutions. Routledge, pp 77–90
Ngandu KY (2017) The panel of the wise: its role in preventing violent conflicts in Africa. UN Chronicle. https://www.un.org/en/chronicle/article/panel-wiseits-role-preventing-violent-conflicts-africa
Okthariza N (2019) The Papua question for Jokowi's second term. East Asia Forum. https://www.eastasiaforum.org/2019/11/16/the-papua-question-for-jokowis-second-term/
Ondawame O (2013) Statement made by dr. otto Ondawame, vice chairman of West Papua national coalition for liberation during the plenary session of MSG meeting in Noumea, Kanaky-new Caledonia. [In-Person]. 19th MSG-Leaders-Summit. https://www.msgsec.info/wp-content/uploads/Statements/2013-Jun-21-19th-MSG-Leaders-Summit-Statement-by-VC-of-WPCL-Dr.-Otto-Ondawame.pdf
Osborne R (1985) Indonesia's secret war: the guerilla struggle in Irian Jaya. Allen & Unwin
Ozcelik S (2006) The theory and practice of preventive diplomacy: the case of preventive deployment in Macedonia. Uluslararası İlişkiler 3(11)
Parameswaran P (2016) The truth about Thailand's 2017 defense budget "Hike." Diplomat. https://thediplomat.com/2016/09/the-truth-about-thailands-2017-defense-budget-hike/
Philpott S (2018) This stillness, this lack of incident: making conflict visible in West Papua. Crit Asian Stud 50(2):259–277
Poling G (2013) Are Pacific Nations Helping the West Papuan Case for Greater Autonomy? Pacific Partners Outlook. Centre for Strategic and International Studies. https://www.csis.org/analysis/pacific-partners-outlook-are-pacific-nations-helping-west-papuan-case-greater-autonomy#commentary
Porto JG, Ngandu KY (2015) The African Union's Panel of the Wise: a concise history. The African Centre for the Constructive Resolution of Disputes (ACCORD)
Radio New Zealand (2016) Melanesian free trade agreement to start in 2017. Radio New Zealand. https://www.rnz.co.nz/international/pacific-news/304902/melanesian-free-trade-agreement-to-start-in-2017
Radio New Zealand (2019) Africa Caribbean pacific group seeks action on Papua rights abuses. Radio New Zealand. https://www.rnz.co.nz/international/pacific-news/405595/africa-caribbean-pacific-group-seeks-action-on-papua-rights-abuses
Saltford J (2003) The United Nations and the Indonesian takeover of West Papua, 1962–1969. Routledge
Scott C, Tebay N (2005) The West Papua conflict and its consequences for the Island of New Guinea: root causes and the campaign for Papua, land of peace. The Round Table 94(382):599–612
Shelton T, Wibawa T (2019) Why West Papua's latest protests are different from the ones before. ABC News. https://www.abc.net.au/news/2019-09-04/west-papua-latest-protest-over-referendum-for-independence/11471016
Singh B (2017) Papua: geopolitics and the quest for nationhood. Routledge
Solomon Times (2016) Allegations of human rights abuse in West Papua Raised at the UN—Solomon times online. Solomon Times. https://www.solomontimes.com/news/allegations-of-human-rights-abuse-in-west-papua-raised-at-the-un/8448

Stott DA (2011) Indonesian Colonisation, resource Plunder and West Papuan Grievances. Asia-Pacific J 9(12):1–33
Sundaryani FS (2017) Papua, West Papua need better connectivity: Jokowi. The Jakarta Post. https://www.thejakartapost.com/news/2017/07/20/papua-west-papua-need-better-connectivity-jokowi.html
Suriadireja K (1987) Pengembangan Irian Jaya Ditinjau dari Segi Ketahanan Nasional. In: Kasiepo M, SK I, Razak A (eds) Pembangunan Masyarakat Pedalaman Irian Jaya. Pustaka Sinar Harapan
Tasevski O (2022) Indonesia's new cabinet and its human rights implications | Lowy Institute. The Interpreter. https://www.lowyinstitute.org/the-interpreter/indonesia-s-new-cabinet-and-its-human-rights-implications
Tebay N (2005) West Papua: the struggle for peace with justice. Catholic Institute for International Relations (CIIR)
Vatikiotis M (2009) Southeast Asia's internal conflicts: regional mediation tools. Council for Security Cooperation in the Asia Pacific, pp 16–21
Webb-Gannon C, Elmslie J (2014) MSG headache, West Papuan Heartache? Indonesia's Melanesian Foray. Asia-Pacific J 12(47)
West Papua Daily (2022) President has visited Papua 14 times affirming Papua is special. West Papua Daily. https://westpapuadaily.com/president-has-visited-papua-14-times-affirming-papua-is-special.html
Wilson C (2001) Internal conflict in Indonesia: causes, symptoms and sustainable resolution. Parliamentary Library of Australia
Wirz S (2013) Asia's Palestine? West Papua's independence struggle. The Diplomat. https://thediplomat.com/2013/11/asias-palestine-west-papuas-independence-struggle/
Wing J, King P (2005) Genocide in West Papua? The role of the Indonesian state apparatus and a current needs assessment of the Papuan people. Centre for Peace and Conflict Studies, & ELSHAM Jayapura
Yabi GO (2010) The role of ECOWAS in managing political crisis and conflict. Friedrich-Ebert-Stiftung, pp 1–57
Yan S, Guo Y (2020) Origin and development of preventive diplomacy. In: Preventive diplomacy in the Asia-Pacific. World Scientific, pp 1–344
Yapumi C (2016) MSG chairs say West Papua will be admitted. Loop PNG. https://www.looppng.com/content/msg-chairs-say-west-papua-will-be-admitted
Zyck SA, Muggah R (2012) Preventive diplomacy and conflict prevention: obstacles and opportunities. Stab: Int J Secur Dev 1(1):68

Strategic Environmental Assessment of Vientiane Transport Master Plan Study Under COVID-19 Pandemic Situation: Meaningful Information Disclosure/Social Acceptance and Its Establishment of Procedural Justice

Thodsakhone Razmountry, Takanori Hayashida, and Souphany Heuangkeo

Abstract Proper information disclosure, appropriate public participation, and building consensus among many stakeholders are crucial to the formulation of a successful Master Plan in order to achieve a stabilized and peaceful urban and/or regional development. Vientiane Transport Master Plan (VTMP) considers the sustainable urban transport system under the vision of inclusiveness with the target year of 2040, to ensure the sustainable urban and transport system. A participatory strategic environmental assessment (SEA) scheme is developed within the VTMP formulation study. Although both VTMP and its SEA are still in progress, it is found that the developed participatory SEA approach is effective and useful in identifying and addressing transboundary issues associated with the urban planning process. Also, it is found that a developed SEA approach supports the smooth consensus building across various stakeholders to guarantee comprehensive urban sustainability. It will also enable obtaining new insights and ideas through interactions among all line ministries and stakeholders and setting the possible ways to achieve a more well-balanced planning process.

Keywords Urban sustainability · Cooperation and conflict over transboundary resources · Land/terrestrial resources planning and management · Conflict and COVID-19

T. Razmountry (✉)
Vientiane Transport Master Plan Study Office, Vientiane Capital, Laos
e-mail: thotsakhone@gmail.com

T. Hayashida
International Development Center of Japan, Tokyo, Japan
e-mail: hayashida.t@idcj.or.jp

S. Heuangkeo
Ministry of Public Works and Transport, Government of Lao PDR, Vientiane Capital, Laos

A. Sharifi et al. (eds.), *Bridging Peace and Sustainability Amidst Global Transformations*, World Sustainability Series,
https://doi.org/10.1007/978-981-99-7572-3_8

1 Introduction

1.1 Consensus Building Through Meaningful Information Disclosure, Empowerment of Communities/Stakeholder

The Master plan (MP) study for the middle and long-term regional/urban development shall consider many factors while managing many opinions of various stakeholders for its successful and sustainable MP implementation. In general, this process takes a lot of efforts to seek a unanimous agreement among all stakeholders in order to avoid potential conflicts that may be triggered due to the lack of the communication and/or miscommunications, but its outcome would be invaluable. Throughout these processes, mentioned above, a successful consensus building would be achieved while building the trust among all stakeholders, and eventually would create a strong, positive MP ownership for all stakeholders. Eventually, it would induce the autonomous commitment from all stakeholders during its future MP implementation phase. Consequently, a formulated MP would be robust and sustainable for middle-term and/or long-term period, and would lead to stabilized, peaceful human society.

Similarly, within Vientiane Transport Master Plan (VTMP) development project, an effective platform shall be developed in order to provide the opportunity to all stakeholders to access VTMP-related information and then, to establish smooth consensus among therein by a proper information disclosure and appropriate public involvement processes. Meanwhile, this consensus building process expects to enhance the liaison among all stakeholders, creates a common understanding of issues and the prospective, to be shared through a series of discussion, that would bridge differences and would generate to higher-quality solutions for existing urban issues. Using this platform, some communities of Vientiane would be able to shape their future urban environment jointly and then advance further with confidences.

Besides, keeping the procedural justice (e.g., Lind and Tyler 1995) throughout the entire MP formulation process is one of another important factors for the successful and sustainable implementation of a middle/long-term infrastructure development MP. Many approaches for the meaningful consensus building have been developed and implemented within current urban development MP studies. One of recent those approaches are the implementation of the participatory SEA within the urban development master plan studies (JICA 2014, 2017), in that a series of intensive stakeholder meeting and workshop with relevant information disclosure while establishing the procedural justice were conducted.

Besides, the implementation of relevant civic education programs are important to deepen stakeholder's understanding of MP. Within VTMP-SEA, this program was implemented to provide learning opportunities of VTMP formulation to stakeholders and to overcome the asymmetry of the MP-related information between the MP planner (i.e., government) and VTMP stakeholders as much as possible.

Inclusiveness is also one of important keywords of VTMP study in order to address various stakeholder groups. It is essential to invite any stakeholders who may have interests/concerns about MP study into a participatory MP development process, that would affect their social activities and/or living conditions in future; and to encourage them to participate this MP development process through this VTMP formulation process.

Expected results of this VTMP-SEA would be the formulation of VTMP that would be acceptable for all stakeholder groups and people become eager to get involved within the updating and/or follow up of VTMP study in more constructive ways. It is noted that all citizens are qualified to be stakeholder of this VTMP since livelihoods/livings of all citizens have close relationship with the current/future urban transport system of Vientiane to some extents.

To make this interaction more active and visible, an ILEK (integrated local environmental knowledge, (e.g., Sato and Kikuchi 2018))-based survey, to be described later, is conducted how each stakeholder think about the current/future Vientiane as well as VTMP. Advantage of this survey is to raise the awareness of the importance of MP development among each stakeholder through this survey, and to get them familiar with current/future urban issues of Vientiane. Using this survey result, it is also possible to create visible images how each stakeholder think about the current/future of Vientiane, to establish platform of easy access for possible future discussions of VTMP among stakeholders and to encourage them to grow their conscientiousness as the MP owner of VTMP since they already live therein.

1.2 Procedural Justice

The concept of the procedural justice grasps at both individual and institutional level that would allow the interlinkage of the inside and the outside stakeholders to have an opportunity to emphasize their concerns and be part of development process by telling their experiences and/or concerns. The concept of the fairness in procedures for solving disputes and allocating resources is known as the procedural justice. Its concept can be used in non legal cirsumstanes when certain procedures are used to settle disputes or split rewards and liabilities.

The distibtive justice (such as the fairness in the distrinbution of rights or resources) and the retributive justice (such as the fiarnesss in the punishment of wrongs) can be contrasted with the procedural justice, which is concerned with the fairness and transparency of processes by which decisions are made. Contacting and hearing all parties and/or stakeholders before the ruling of a decision, is one appropriate step to be taken in order to make a process procedurally fair. Some of the procedural justice theories say that a fair procedure would lead to fair outcomes, even if the requirements of a distributive or a restorative justice are not matched. It has been explained that this is the outcome of the higher quality inter-personal interactions, sometime observed within the procedural justice process, which has indicated to be stronger in affecting the perception of the fairness during the conflict and/or dispute resolution process. Within VTMP, the procedural justice is expected to be established thorugh following steps, i.e., (i) a series of interactions among city-wide stakeholders and the MP planner, (ii) meaningful and effective information disclosure, (iii) the empowerment of stakeholders through the participation into the partial MP development (e.g., establishment of ILEK for VTMP among all stakeholders, and (iv) a step-by-step concensus building through processes, mentioned above.

1.3 Vientiane Transportation Master Plan (VTMP)

The Ministry of Public Works and Transportation (MPWT), the Government of Lao PDR, with the financial and technical supports from Japan International Co-operation Agency (JICA), is currently implementing the VTMP formulation study. Aims of this study is to upgrade the urban infrastructure, make it accessible to all people in Vientiane through the improvement of the urban public transportation quality as well as to promote in use of the public transportation system, that would lead to the reduction of the road accident and the air pollution.

Currently, many urban transport and relevant development projects such as the High-Speed Railway, Lao-Thai Railway Extension, Inter-provincial Highway connecting Vientiane and Vang Vieng, Bus Rapid-Transit (BRT), City Bus, relevant urban re-development/new township, and others are implemented across Vientiane Capital without a comprehensive coordination and/or harmonization. Besides, a railway project originating from the existing High-Speed Railway Station to Vietnam and the Cable Car crossing the Mekong River are at the planning stage.

Both JICA and the Government of Lao PDR agreed that Public Works and Transport (PWT) Sector, JICA's counterparts, would be in charge of implementing this VTMP project. They also agreed that PWT sector would coordinate with other pertinent organizations and make sure that the project would continue to operate independently both during and after the implementation period in order to support Lao PDR's socio and economic development. The Followings are the list of the project implementing/counterpart (C/P) agencies (Table 1).

Figure 1 shows the area of concerns for this VTMP study. Originally the study period of this VTMP project was of three (3) years, initiating from December 2018. Due to the COVID-19 pandemic and relevant travel restrictions, some of the project activities have been delayed. Thus, in order to complete all project activities, the entire project duration is extended to September 2023.

Table 1 List of counterpart agencies as project implementing agencies

–Department of Transport (DoT), MPWT
–Public Works and Transport Institute (PTI), MPWT
–Department of Public Works and Transport (DPWT) at Provincial Level, Vientiane Capital
–Department of Traffic Police at Provincial Level (DTP), Vientiane Capital
–Vientiane Capital State Bus Enterprise (VCSBE)
–Steering Committee to Solve Congestion and Traffic-Management in Traffic Routes in the City of Vientiane Capital (CTMC)

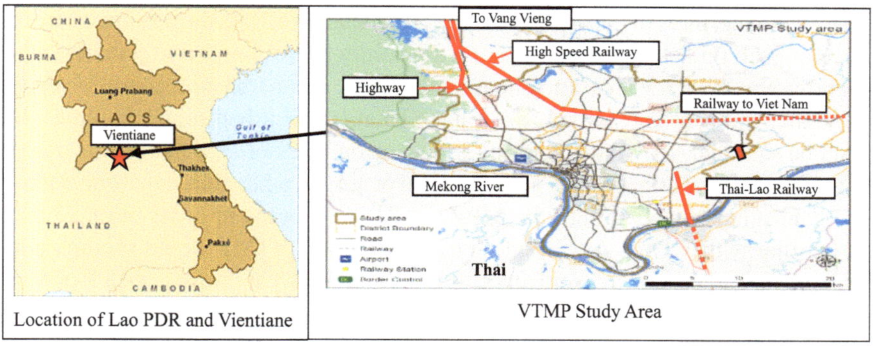

Fig. 1 Study area. 2015 *Source* After PowerPoint file, used within the presentation of Asia Impact Conference 2022 (Razmountry and Hayashida 2022). *Note* Population of Vientiane is of 820,940 people (Lao Statistic Bureau). *Source* https://search.yahoo.co.jp/image

1.4 Strategic Environmental Assessment (SEA)

1.4.1 Trans-Disciplinary (TD) Approach

The merging of several disciplines has always benefited the urban planning study. The rudimentary urban planning began to take shape in the late 19th century as a result of interactions among professionals in civil engineering, architecture, medicine, public health, sociology and, occasionally biology. Many urban research programs were reorganized in the 1960s under the umbrella of urban studies, incorporating even more social disciplines including sociologies, economics, geography and political science.

In general, an integrated urban planning approach embraces the ambiguity to enhance cities' abilities to be liveable, resilient and the effective. It would collaborate with architects, urban planners, waste managers, energy providers, transport planners and other relevant experts rather than ignoring various viewpoints. TD approach is expected to identify solutions for many urban-related issues holistically at the regional scale (Sato and Kikuchi 2018), and SEA facilitates a transboundary co-operation by sharing all development-related information and providing a common platform of the dialogue among all stakeholders. Eventually, this would enhance the solution finding process under the environmental resources of concerns, and then, would contribute to the conflict prevention.

SEA-based TD approach tools for a collective discussion are developed in order to strengthen the motivation to learn VTMP as well as the townscape of Vientiane, and eventually to empower its understanding among all stakeholder while addressing urban issues to be prioritized and/or local knowledge each stakeholder owns. Eventually, this approach would collect and then, combine those issues and/or knowledge, that would play vital role for creating a sustainable urban transport system and for implementing relevant urban development activities and/or plans.

In Vientiane, most of departments at the provincial level (i.e., the provincial branch office of each ministry) develop their own medium and/or long-term policies, plans and programs (PPP). VTMP encompasses the future urban transport condition across Vientiane, and can be regarded as a fundamental for all urban social activities. So, it can be said that VTMP has a strong influence on PPPs of all departments. As the first step of this TD approach, it is imperative to get other line departments and/or organizations involved into the MP formulation process in order to make them get familiar with VTMP. As a result, it is expected that VTMP would become more harmonized and be able to avoid any confrontations of VTMP with any PPP of those line departments.

1.4.2 SEA

SEA method mainstreams and advances the environmental sustainability in the decision-making process. SEA of PPP strives to integrate environmental factors into PPP development and assess how they interact with economic and social factors (Ninomiya et al. 2016). The Organization for Economic Cooperation and Development (OECD)'s Development Assistance Committee summarizes following benefits, to be caused by applying SEA to the development cooperation (Table 2).

Environmental mainstreaming began at the project level as the Environmental Impact Assessment (EIA) within the framework of the development aid. EIAs are meant to enhance the project design and its execution by finding methods to avoid, reduce and compensate for any negative environmental consequences during the feasibility study stage.

However, due to inadequate interaction and a disorganized relevant information sharing system across competent governmental agencies and/or entities in certain countries, the development of project and/or plan consensus has often been difficult, resulting in the disagreement among key stakeholders as well as large-scale negative impact on both natural and social environment.

To address those issues, there is growing acknowledgement of the importance of considering environmental consequences of regional and sectoral development plans at the macro level. The SEA's goal is to include environmental and social factors

Table 2 Significant of strategic environmental assessment

• Enable better informed decision-making by providing the environmental data
• Encourage an organized, comprehensive assessment of development possibilities to find new opportunities
• Prevent expensive errors by letting decision-makers know about potentially unsustainable development alternatives at an early stage
• Increase stakeholder participation in decision-making for better governance
• Protect environmental assets for long term development and poverty reduction
• Facilitate trans-boundary co-operation around common natural resources and assist minimize conflict

into the formulation process of PPPs at the early planning stage, mitigate negative implications and maximize possible benefits synergies by encouraging information exchange and dialogues among key parties.

1.5 Participatory SEA of VTMP

Within VTMP study, SEA is employed from the very beginning stage of PPP formulation in order to analyze its potential effectiveness. A participatory SEA implementation framework was developed aiming to examine how economic and social factors interact with environmental factors in order to (i) make VTMP acceptable not only to line ministries/agencies and/or organizations but also to local communities of Vientiane through a series of information disclosures and dialogues while establishing a common discussion platform with relevant procedural justice under COVID-19 pandemic situation, and (ii) to achieve a smooth consensus of MP formulation through a series of step, mentioned earlier. Table 3 summarizes major activities of this participatory SEA of VTMP study (VTMP-SEA).

As described later, SEA is relatively new institutional system in Lao PDR, and this VTMP-SEA is the fourth SEA implementation practice in Lao PDR. As of July 2022, no SEA practice of PPP development projects is completed yet (MONRE, personal communication 2022). Before starting VTMP-SEA officially, a series of the capacity development program was conducted since September 2021. More detailed description of this SEA-related capacity development is to be summarized later. It is noted that this VTMP-SEA is still on-going, and to be finalized around September 2023.

Due to the COVID-19 pandemic situation of Vientiane around the 1st and 2nd quarter of Year 2022, an implementation of large-scale, direct (face-to-face) meetings were banned. Instead, the use of the social media with newspaper advertisements were implemented in order to assist appropriate information disclosure, announcements of VTMP-SEA as well as to establish a dialogue platform for enhancing relevant question and answer session.

During the scoping study phase of this VTMP-SEA (see Table 3), several organizations that would have relatively large numbers of the membership and be influential on communities across Vientiane were contacted in order to have an off-line, small-sized but effective SEA-related workshops. Mainly these workshops consist of focal and resources persons and it is expected to spread relevant information of VTMP-SEA to the community-level using internal liaison mechanisms of each organization. As a result, three organizations such as Lao Women Union (LWU), Lao People's Revolutionary Youth Union (LYU) and Lao Small-Medium Enterprise (SME) (see Table 3), were selected within VTMP-SEA.

Table 3 Major activities for VTMP-SEA implementation

Process	Schedule	Core participants
Preparation of VTMP-SEA	July/2021–Dec/2021	Implement a series of SEA-related capacity building for MONRE, MPWT, DoT, DPWT, NUOL and others
Scoping study of VTMP-SEA	Jan/2022–April/2022	Kick-off meetings with key stakeholders such as counterpart agencies/PWT sectors, environmental authorities (MONRE/DONRE), Urban management authority (Vientiane capital office for management and service)
Multi-tier workshop/group-wide stakeholder meeting	March/2022–June/2022	Internal VTMP-SEA related awareness seminar inside of PWT sector (MPWT/DPWT/DoT)
		To create the city-wide awareness and obtain the inputs for VTMP implementation –Lao women union (LWU) –Lao people's revolutionary youth union (LYU) –Lao small-medium enterprise (SME)
		Inter-ministry at the regional level (39 Departments) to disseminate VTMP-related info while establishing MP consensus
Overall stakeholder meeting	Aug/2022	1st general stakeholder meeting
	Dec/2022	2nd general stakeholder meeting
Submission of VTMP-SEA draft final report	Feb/2023	SEA draft final report was submitted from DPWT to DONRE
Validation meeting	Sep/2023 (tentative)	Disclose relevant information of VTMP (final draft) and relevant SEA study results

Note MONRE, DONRE and NUOL stand for "Ministry of Natural Resources and Environment", "Department of Natural Resources and Environment" and "National University of Laos" respectively

2 Environmental and Social Legal Framework in Lao PDR

In Lao PDR, the Ministry of Natural Resources and Environment (MONRE) is the presiding central governmental body for the environmental management and there are following three SEA-related legal codes, i.e., (i) Environmental Protection Law (No: 29/NA Dated on 18 December 2012, Article 19), (ii) Decision on SEA approved by Minister of MONRE, No: 0483/MONRE, Dated on 06 February 2017, and (iii) SEA Guideline (No: 6616/MONRE, Vientiane Capital City, Dated on 31 December 2018). SEA is a relatively new institutional framework in Lao PDR, and unfamiliar to the most of governmental organizations in Lao PDR (MPWT/DPWT included).

Before that, SEA was practiced at donor-funded MP development projects sporadically around 2010. As mentioned earlier, SEA is relatively new system in Lao PDR, and MONRE plans to revise and improve its contents, reflecting feedbacks from SEA

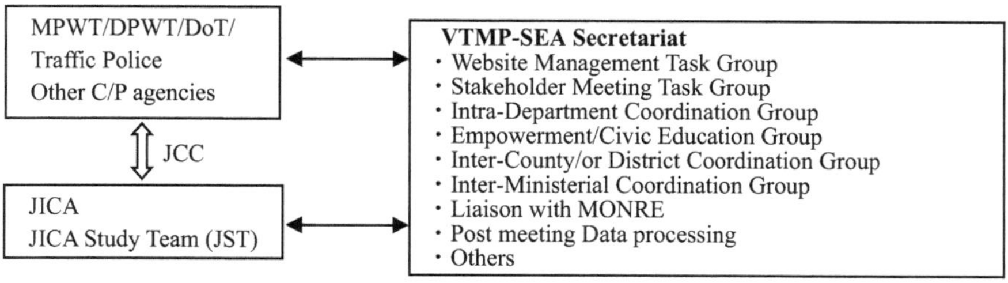

Fig. 2 VTMP-SEA implementation framework. JCC: joint coordination committee, periodically held between JICA and C/P agencies

implementation case studies inside and outside of Lao PDR continuously (MONRE, personal communication 2022).

3 VTMP-SEA Implementation in Lao PDR

3.1 Overall Implementation Framework

To ensure the seamless implementation of this SEA, a new ad-hoc organization, called "Master Plan Secretariat", is established while being supervised by both JICA Study Team (JST) and DPWT (see Fig. 2).

The newly established secretariat's primary responsibility is to oversee all SEA implementation steps with relevant technical co-operation of JST. Also, this secretariat establishes the connection among MONRE/DONRE, relevant governmental bodies, colleges universities, other international institutions as well as JICA and other important stakeholders. This secretariat is in charge of organizing the information disclosure as well as the public relation campaign, one of the crucial elements of this SEA process through the website of VTMP-SEA, to be described later.

3.2 Capacity Development Program for MPWT/DPWT

No environmental unit nor section exist within both MPWT and DPWT, MP owner of VTMP, thus there is no implementation experience of SEA therein. So that, relevant SEA-related capacity development program was established, and then, implemented during both the preparatory phase and the early scoping stage of VTMP-SEA to raise awareness of SEA while stressing out that SEA is one of useful urban/regional planning tools and does not confront to any conventional urban and/or regional development MP study approaches. As of October 2022, eight (8), on-line based SEA-related capacity development seminars were conducted, using current JICA-financed SEA implementation case studies (see Table 4). Also, representatives of

Table 4 Contents of SEA-related capacity development program

	Date	Contents
1	Sep 29, 2021	What is SEA?
2	Oct 06, 2021	Case Study #1: participatory SEA, implemented within port development MP study in Burundi, Africa
3	Oct 13, 2021	Tools for decision making: collective discussion of alternatives in MP study, Mozambique, Africa
4	Oct 20, 2021	Public consultation: Nairobi Urban Development MP, Kenya, Africa
5	Feb 04, 2022	Guidance 1: expected roles of MPWT/DPWT officials within VTMP-SEA
6	Feb 14, 2022	Guidance 2: sector-wide group workshop (MPWT/DPWT) within VTMP-SEA
7	Feb 25, 2022	Guidance 3: sector-wide group workshop (MPWT/DPWT) within VTMP-SEA
8	April 20, 2022	Progress of VTMP-SEA

many sectors such as not only MPWT and DPWT but also MONRE, NUOL and private sectors were invited in order to disseminate SEA-related knowledge while making SEA more familiar in the Lao context.

3.3 *Empowerment of Participants*

Within VTMP-SEA workshops, several empowerment tools such as an ILEK-based brain-storming session, the development of both compatibility and compound matrices were introduced after the explanation of VTMP (see Table 5). The main purpose of these sessions is to make participants get familiar with the usage of some of urban planning tools as well as the prioritization of key urban development issues through these workshop exercises. Key discussion points and/or topics of each session, summarized within Table 5, were pre-prepared by JST. In order to handle the smooth workshop, small groups made up of participants were formed. There were three facilitators including one assistant, who have been trained before the initiating a series of the empowerment workshop in order to support the streamlining discussions of each divided small groups.

Table 5 Empowerment for participants of VTMP-SEA

Session	Purpose and outline
ILEK brain storming session	To develop collective environmental and social knowledge of current Vientiane through the brain storming discussion while precipitating each participant's interaction and deepening the information sharing on the current urban transport situation. Based on those knowledges, then, priority of raised urban development issues are to be addressed
Compound matrix	Way in which PPP will interact with each criterion in matrix is discussed by review group & view is taken as whether PPP is likely to eliminate environmental effect or worsen or be largely neutral
Compatibility matrix	To determine degree to which PPPs support or work against each other—in other words how compatible they are

3.4 Social Media Management

As mentioned earlier, due to the outbreak of COVID-19 in Vientiane around by May 2022, direct face-to-face meetings with more than 50 participants were banned. In order to maintain the information disclosure and to guarantee the accessibility of VTMP and its SEA-related information, the website page for VTMP-SEA is created using Facebook platform (its domain name is https://www.facebook.com/profile.php?id=100083232495591). Primary goals for developing of this website page are to (i) encourage public involvement via cyber-space, (ii) allow possibilities for people of all ages and important stakeholders to learn the creation of MP, (iii) provide general public opportunities for the public involvement in the MP formulation process, and (iv) gather VTMP-related opinions from the general public and important stakeholders. Those feedbacks are to be integrated within the MP formulation process.

4 Discussions

In conjunction with the proposed vision of VTMP 2040, "towards an accessible and sustainable city for everyone in 2040 and proposed mission of create an inclusive, sustainable and modern transport system along with a joyful walkable environment", the stakeholder-consultation is regarded one of the key components of this VTMP-SEA. In general, the stakeholder meeting is conducted at several different timing in order to achieve the consensus of PPP while precipitating the interaction between PPP makers and the general public through appropriate information disclosure and the civic education.

As one of key actors of this VTMP study, PWT Sector (MP owner) shall proceed a series of the stakeholder meeting in collaboration with other line governmental organizations and/or agencies such as the Ministry of Industry and Commerce, the Ministry of Information, Culture and Tourism, the Ministry of Labor and Social

Welfare, MONRE, the Ministry of Planning and Investment, the Ministry of Health, the Ministry of Public Security, Vientiane Municipality and other equivalent organizations. In Lao PDR, no inter-ministry PPP development process was taken in the past. In that aspect, it can be said that this VTMP project is unique and the first case study of the master plan development, using the participatory SEA framework while encouraging other line ministries and/or departments to have constructive interaction. It is noted that 39 departments at the provincial level participated within this on-going SEA study of VTMP project.

In addition, as mentioned earlier, MPWT has several on-going infrastructure development projects such as High-Speed Railway, Thai-Laos Railway Extension, BRT, Inter-provincial highway, City Bus as well as others, and those projects are not well-coordinated one another. So that, it would be essential to set-up a platform of the intra-stakeholder meeting even within PWT Sector such as MPWT, DPWT, PTRI, DTP, and CTMC.

To guarantee the procedural justice under COVID-19 pandemic situation which did not allow to organize large-scale direct (off-line) meetings, several organizations that would have a large membership and an influential, smooth internal spreading effect of the information were contacted (see Table 6). These selected organizations own similar goals with different prospective on contributing the national development and cover the responsibility from localized to centralized levels, which enable to ensure VTMP-SEA-related information disclosure and dissemination is to be taken place effectively. Within this VTMP-SEA, it is found that collaboration with those selected organizations is quite useful and helpful, in particular, within the inclusiveness, the information disclosure, empowerment, constructive discussions and VTMP-related decision making.

As summarized in Table 6, one of main objectives of VTMP-SEA is to guarantee the inclusiveness of all stakeholders under COVID-19 pandemic situation. In general, possible stakeholder groups shall be addressed to share MP-related information, and then invite those into participatory MP development process. Eventually this would affect their future social activities and/or living conditions across Vientiane; and would encourage them to participate this MP formulation process through this VTMP-SEA more constructively. It would be better to study the internal social dynamics of each organization in order to have more effective participatory SEA implementation framework.

A series of workshops, conducted within VTMP-SEA (see Fig. 3) enable both MP owners and participants to raise awareness of future city-wide situation through the urban transport planning, and then recognize the importance of VTMP. For instance, using a common discussion platform, established within this VTMP-SEA, all participants (MP owner included) can understand the way how several urban transport projects are implemented as well as pros and cons of each individual project from the viewpoints of bio-physical, socio-economic and socio-cultural aspects. Consequently, they begin to notice potential urban issues to be addressed with short-term, middle-term and long-term perspectives by sharing common understanding and knowledge through the constructive interaction and/or discussion processes. Eventually, this would lead a collaborative PPP formulation, achieving the common

Table 6 Selective sector-wide stakeholders

Organization (year of establishment)	Main activities and key features	Total number of membership	Key organization structures (area of responsibility)
LWU (1955)	–To empower and gather Lao women to fully participant in the process of the national development and preservation –To encourage the engagement of women in the process of implementing the country's socio-economic development –To promote the empowerment of women from all ethnic and social background by providing a comprehensive of the government's policies, constitution, laws and numerous international instruments. By equipping women with this knowledge, the foster active participation and engagement in the democratic processes that shape the society can be expected	1,015,506 women (as of 2009)	Throughout the country at four levels, as central; provincial/ministerial, district/municipal and village ones
LYU (1955)	–To motivate young people throughout the country to support national development by focusing on the field of information, media, entertainment, art and music	243,500 registered number (aged 15–35)	Central, provincial, municipal, district and village levels and cooperates with foreign countries and international organizations in a wide range of program activities
Lao SME (2006)	Under the wing of the Ministry of Industry and Commerce –To facilitate international business relations with production and trade mechanism and sustaining and creating local job employment	124,567 enterprises	3 major business sectors (i.e., production of commercial goods, trade and service sector) Average no. of employee Micro enterprise: 1–5 Small enterprise: 6–50 Medium enterprise 51–99

Source Promotion and Protection of Lao Women's Rights, Lao Women's Union, 2009
Source Lao Youth Union
Source ADB Asia SME Monitor 2020 database. Data from the Ministry of Industry and Commerce; Lao National Chamber of Commerce and Industry

goals of the sustainability by reflecting the characteristic of local communities and addressing specific needs of each community.

Discordances/conflicts and/or disputes on the planning process tend to arise from the lack of a proper information disclosure and the shortage of relevant communication and/or dialogue among stakeholders. In the beginning of the workshop, participants sensed unfamiliarity with the planning tools, namely, compound and compatibility matrices, mentioned earlier. However, as their understandings of VTMP started to deepen, an entire discussion such as the development of ILEK as well as prioritized urban issues, associated with the development of VTMP became active and streamlined soon.

Under the COVID-19 pandemic circumstance, the information disclosure shall not be limited but rather addressed to the general public via a social media platform, namely, Facebook page, to draw more attentions from the wide-range of community across Vientiane. Figure 4 shows the access trend of VTMP-SEA during last 4 months (May 2022–August 2022). From this figure, it can be seen that the total access number of VTMP-SEA Facebook tend to increase gradually. During this period, there were several VTMP-SEA-related events (see Table 3). That may cause stimulus and raise some concerns of VTMP among internet users of Vientiane. It is essential to monitor this trend, collect more relevant data, analyze more long-term trend of this type of social media activities, and then, evaluate the usefulness.

Figure 5 shows the questions and answer session, conducted within a developed VTMP-SEA Facebook. As of August 2022, 8 (eight) comments regarding the VTMP project were delivered to DPWT, and then, replied to each by DPWT (i.e., MP owner) later therein. Based on the social media report (as of August 2022), it is found that this VTMP-SEA Facebook are accessed not only from the citizen of Vientiane but also from other provinces of Lao PDR as well as other countries.

5 Conclusions

The fundamental benefit of SEA is to integrate environmental and social safeguards into PPP formulation. As a result of engagement of all stakeholders, the VTMP shall be validated by appropriate information sharing, the public involvement and strong consensus buildings through a series of constructive discussion. A participatory SEA framework was developed and then, was implemented to strengthen the liaison among PWT Sector (i.e., MP owner), line agencies, and various stakeholders while paying attention to the establishment of the procedural justice. It is important to organize multiple workshops and/or stakeholder meetings with same participants in order to deepen their understandings of VTMP and to intensify their interactions with the future Vientiane through the implementation of VTMP. Although, both VTMP and its SEA studies are currently in progress, PWT Sector recognized the usefulness of SEA within the PPP planning process. SEA enable them to obtain new insights

a. MPWT/DPWT Internal VTMP-SEA related awareness seminar, held on March 23rd, 2022 (part 1)

b. MPWT/DPWT Internal VTMP-SEA related awareness seminar held on March 23rd, 2022 (part 2), Group Discussion

c. MPWT/DPWT Internal VTMP-SEA related awareness seminar held on March 23rd, 2022 (part 3), Group Discussion

d. Group-wide VTMP-SEA Workshop at LWU, held on April 05, 2022 (part 1), Explanation of VTMP

e. Group-wide VTMP-SEA Workshop at LWU, held on April 05, 2022 (part 2), Group Discussion, the person left-below is a facilitator.

f. Group-wide VTMP-SEA Workshop at LWU, held on April 05, 2022 (part 3), Presentation and finding-sharing of Group Discussion Results.

g. Group-wide VTMP-SEA Workshop at LYU, held on April 21, 2022 (part 1), Question and Answer Session regarding the explanation of VTMP

h. Group-wide VTMP-SEA Workshop at LYU, held on April 21, 2022 (part 2), Group Discussion

i. Group-wide VTMP-SEA Workshop at LYU, held on April 21, 2022 (part 3), Presentation and finding-sharing of Group Discussion Results.

j. Group-wide VTMP-SEA Workshop at Lao SME, held on April 28, 2022 (part 1), Explanation of VTMP

k. Group-wide VTMP-SEA Workshop at Lao SME, held on April 28, 2022 (part 2), Group Discussion

l. Group-wide VTMP-SEA Workshop at Lao SME, held on April 28, 2022 (part 3), Presentation and finding-sharing of Group Discussion Results.

m. Inter-Ministry Workshop at Provincial Level VTMP-SEA related awareness seminar, held on June 9th, 2022 (part 1). Representatives from 39 departments attended.

n. Inter-Ministry Workshop at Provincial Level VTMP-SEA, held on June 9th, 2022 (part 2), Group Work and Discussion

o. Inter-Ministry Workshop at Provincial Level VTMP-SEA, held on June 9th, 2022 (part 3), Presentation and finding-sharing of Group Work & Discussion Results.

Fig. 3 Photo records of VTMP-SEA sector-wide workshops. Note that **a–c** VTMP-SEA awareness seminar for MPWT/DPWT, **d–f** working group for LWU, **g–i** working group for LYU, **j–l** working group for Lao SME, **m–o** workshop for inter-ministry at the provincial level

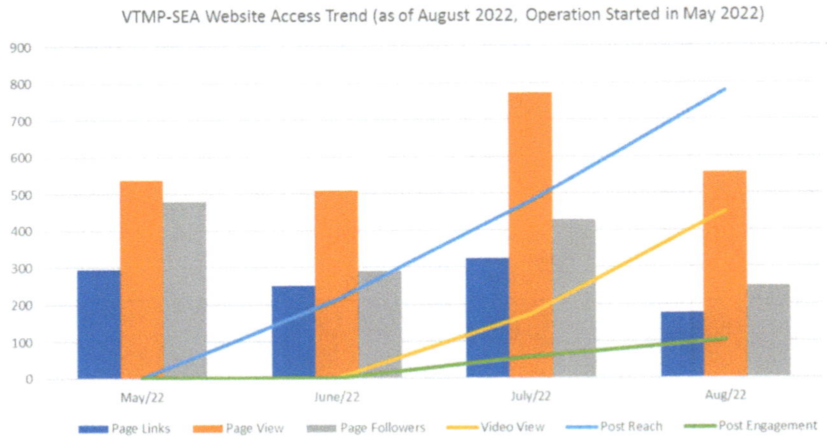

Fig. 4 Time variation of VTM-SEA Facebook access (May 2022–Aug/2022). *Note* Operation of Facebook was initiated in May 2022

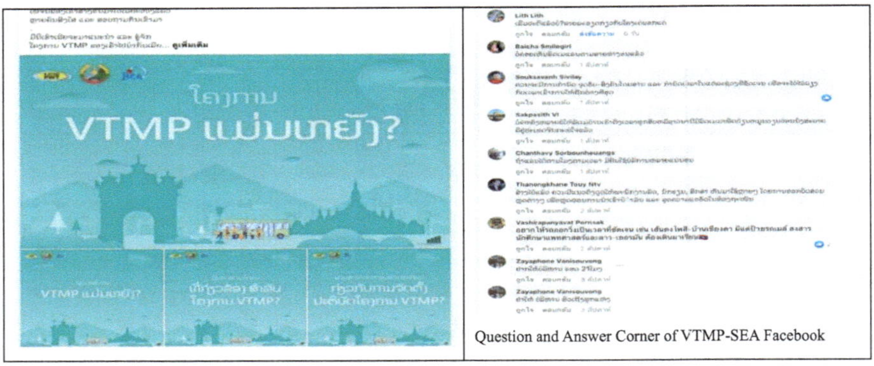

Fig. 5 VTMP-SEA question and answer window (as of August 2022)

and ideas through interactions and setting the possible ways to achieve more well-balanced planning. Consequently, formulated MP would be sustainable for the long-term period, and then, would lead to stabilized, peaceful human society. It is essential to conduct medium- and/or long-term monitoring/follow-up of this on-going VTMP-SEA in order to have accurate evaluations of achievements and/or products, to be obtained through its implementation.

Acknowledgements VTMP study and relevant SEA study are parts of the project, entitled "The Project for Institutional Capacity Building for Sustainable Urban Transport System", financed by Japan International Co-operation Agency (JICA)

References

JICA (2014) Final Report, entitled "The project on integrated urban development master plan for the Republic of Kenya"

JICA (2017) Final Report, entitled "Project for urban development master plan for Managua city in Republic of Nicaragua"

Lao Statistic Bureau (2015) Population Census

Lind EA, Tyler TR (1995) (translation into Japanese by Ohbuchi K.), The social psychology of procedural justice. Brain Press, Tokyo, Japan

MONRE, personal communication (2022)

Ninomiya K, Hayashida T, Mori T (2016) Good strategic environmental assessment (SEA) practices in developing nations: based on Japan's support experiences in Eastern and Southern Africa, Yamanashi global studies. Bulletin of Faculty of Global Policy Management and Communications, vol 11. Yamanashi Prefectural University, Japan, pp 22–38

Razmountry T, Hayashida T (2022) Development of participatory strategic environmental assessment (SEA) Framework for the Vientiane Urban Transport Master Plan (VTMP) updating study in Lao PDR under COVID-19 pandemic situation: Stakeholder analysis of VTMP and Future Urban Development of Vientiane, Proceeding of the 15th Asia Impact Assessment Online Conference (AIC2022), Matsue, Japan

Sato T, Kikuchi N (2018) Transdisciplinary local environment studies (Japanese). University of Tokyo Press, Tokyo, Japan

Disentangling Environmental Justice Dimensions of Urban Green Spaces in Cities of the Global South

Ngoc Thuy Nguyen and Lam Nguyen

Abstract Environmental justice is concerned with the inequitable distribution of environmental resources and effects and encourages public participation in environmental decision-making processes. This concept has been integrated in urban green space planning and design to ensure that urban green spaces can deliver environmental benefits and improve people's wellbeing without exacerbating social injustice. While there have been many studies looking at environmental justice insinuations of urban green spaces in cities of the Global North, not many have focused on cases in the Global South. This research aims to address this research gap by reviewing and synthesizing the body of literature reporting on urban green spaces in cities of the Global South and analyzing how different dimensions of environmental justice have been explored and articulated. This paper also attempts to explore what salient sustainability issues that urban green spaces could address most articles are concerned with, as well as the predisposing factors recognized to predict disparities in green space access. A total of 121 articles published between 2012 and 2022 were thoroughly screened and scrutinized. The findings demonstrate that most articles focus on the distributional justice dimension, while not much attention is paid to the procedural and recognition dimensions. Based on the results, a conceptual framework advising on principal criteria that can lead to inclusive and sustainable green spaces has been developed. This framework links the concept of environmental justice with Goal 11 and Goal 16 of the Sustainable Development Goals and emphasizes the importance of promoting public participation and advocating for transparent and inclusive decision-making processes.

Keywords Green space · Urban planning · Social equity

N. T. Nguyen (✉)
Centre for Environmental Policy, Imperial College London, London, UK
e-mail: n.nguyen21@imperial.ac.uk

L. Nguyen
Urban and Economic Newspaper, Hanoi, Vietnam

A. Sharifi et al. (eds.), *Bridging Peace and Sustainability Amidst Global Transformations*, World Sustainability Series,
https://doi.org/10.1007/978-981-99-7572-3_9

1 Introduction

Despite being considered the key drivers of economic growth, cities are facing many challenges, including urban sprawl, environmental threats, social inequities, and dearth of natural resources. Addressing all these urban challenges and creating a safe, resilient, and inclusive living environment that can enhance the health and wellbeing of urban dwellers are crucial to ensure the sustainable development of cities. The role of nature-based solutions such as green spaces has been widely acknowledged as effective strategies to address environmental and social issues emerging in urban areas (Antuna-Rozado et al. 2019; Swanwick et al. 1978; Wickenberg et al. 2021).

Urban green spaces are public or private urban spaces that have natural components such as vegetation, water features, and so on (Braquinho et al. 2015). There are many types of green spaces, with some examples being urban forests, street trees, green roofs, green walls, parks, and gardens. Urban green spaces have been proven to deliver a wide range of ecosystem services to urban dwellers and important contributors to the sustainable development of urban areas (Antuna-Rozado et al. 2019; United Nations 2015). In fact, urban green spaces contribute to environmental sustainability by regulating thermal temperatures, improving the air quality, alleviating climate change impacts, and so on (Lee et al. 2015; Mayor of London 2018; Schäffler and Swilling 2013; van den Berg et al. 2015). The effects of urban greening on social sustainability, including improving people's mental and physical health (Akpinar 2016; Lee et al. 2015; Nguyen et al. 2021; World Health Organization 2016) and facilitating social cohesion have also been recognized (Ferguson et al. 2018; Klein et al. 2021; Zhang et al. 2021). Urban green spaces have become a prominent research topic and have been advocated in various international policy agendas and guidelines, including the 2013 European Union Strategy and the New Urban Agenda. Most notably, the Sustainable Development Goal (SDG) 11 of the UN 2030 Agenda for Sustainable Development has accentuated the cruciality of providing "safe, inclusive, and accessible green and public space" (United Nations 2015).

Nonetheless, many studies have found that urban greening can perpetuate the inherent social disparities existing in urban areas and usually result in trade-offs between environmental and social development goals in cities. Although green spaces can offer many benefits, they might catalyze gentrification by increasing land value and cause displacement of poor, vulnerable populations (Cole et al. 2017; Curran and Hamilton 2012; Wu and Rowe 2022), thus undermining sustainable development. Many studies have also reported that high-quality green spaces are located near the affluent (Engelberg et al. 2016; Mears et al. 2019; van Dillen et al. 2012; Xu et al. 2017), meaning that the benefits they confer are not proportionately distributed across the populations. When the distribution of a public good is inequitable and differentiated based on socio-economic attributes such as gender, age, income, education, and so on, it deepens social inequity and reduces social sustainability, a pivotal pillar of sustainable development.

If environmental justice considerations are not integrated in urban planning processes, the capacity of urban green spaces in augmenting the social capital of the urban areas may be diminished (de Vries et al. 2020; Kronenberg et al. 2020; Rigolon et al. 2018a, b; Smiley et al. 2016). It is important to understand critically how green spaces are contributing to the community, who can access or benefit from green spaces and with what effects. The focus needs to centered around the impacts of green spaces on different social groups within an urban community, particularly groups that have been previously identified in the literature as having low access to green spaces such as children (Huang et al. 2020), the elderly (Artmann et al. 2019; Gong et al. 2016; Wen et al. 2020), low-income people (Huang et al. 2020; Jones et al. 2009), women (Dadvand et al. 2014), and the racial/ethnic minority people (Byrne and Wolch 2009; Engelberg et al. 2016; Venter et al. 2020).

Although the issue of green space inequities has been extensively explored in studies focusing on cities of the Global North, it is still an emerging topic for cities in the Global South. Importantly, rapid urbanization coupled with an increase in urban population have deepened socioeconomic disparities and exerted disproportionate environmental and economic burdens on people living in the Global South (Arku and Marais 2021). Studies have shown that green spaces could contribute positively to the Global South cities (Feng et al. 2019; Hwang et al. 2020; Rigolon et al. 2018a, b), but more attention should be paid to understand the environmental justice insinuations of urban green spaces to ensure the potential of urban green spaces in fostering sustainable development is fully realized, and everyone, regardless of their backgrounds, can benefit from green spaces.

1.1 Current Reviews Linking Environmental Justice and Green Spaces

Social and environmental equities have been an enduring topic in green space research. Although there is one systematic review focusing on the social disparity in green space access in the Global South by exploring the relationship between green space access and personal attributes such as socioeconomic status or ethnic/ racial backgrounds (Rigolon et al. 2018a, b), it has not appraised different environmental justice dimensions. Besides personal attributes, many other determinants result in disparities in green space access, including physical and social structures.

Some systematic reviews have also comprehensively scrutinized the questions of environmental justice in urban greening. Yet, they covered both Global North and Global South countries. Whilst the wide scope allows for the inclusion of a vast amount of information, it makes it challenging to distinguish the trends between Global North and Global South countries. Particularly, the issues of equity in green space access have been well-researched in Global North countries such as the United States, the United Kingdom, and Canada. On the other hand, urban greening is not the priority in the urban planning agenda of many Global South countries and is, therefore, underrepresented in current research. Grouping literature from all countries together might produce results that are more relevant to a certain group of countries. Hence, it is essential to have a systematic review dedicated to understanding the environmental justice issues in urban greening in the Global South to provide pertinent findings and recommendations.

1.2 Aim and Objectives

This research aims to review the existing literature and explore the relationship between urban green spaces and environmental justice. This will be achieved through the following objectives:

- To synthesize the existing evidence on equities in green space exposure and access in Global South countries;
- To investigate which environmental justice dimensions the included studies use to analyze the issue of inequity in urban greening;
- To explore the predisposing factors recognized in the included studies to predict green space access in cities of the Global South; and
- To understand what sustainability issues that green spaces can potentially address are mentioned frequently across all dimensions of environmental justice.

2 Environmental Justice

2.1 The Concept of Environmental Justice

Conventionally, environmental justice refers to the distribution of and exposure to environmental burdens and benefits by different groups. However, this definition is incomplete as it is concerned with only one aspect of justice, which is equitable distribution (Young 1990). Although safeguarding equitable allocation of resources is crucial to achieving justice, there are other reasons why certain populations have better access to or benefit more from resources (Schlosberg 2004).

Many studies have highlighted that environmental justice encompasses three concepts: distributional justice, procedural justice, and participatory justice. Distributional justice concerns the just allocation of resources. Procedural justice highlights the equity of the decision-making processes. Interactional justice recognizes the diversity of values, beliefs, and perspectives in affected individuals (Honneth 2016; Preston 2015; Schlosberg 2004).

2.1.1 Distributional Justice

Distributional justice, or equity in the allocation of environmental ills and benefits, is the most mainstream dimension of environmental justice in the literature. It has been investigated, in the context of green space planning, in respect of socioeconomic stratification (Nasri Roodsari and Hoseini 2021; Sathyakumar et al. 2019; Shanahan et al. 2014) and the distribution of green spaces in terms of proximity, quantity, and quality (Rigolon 2016). Distributional justice is a major concern for national and regional planning policies, and many have guided the construction and development of green spaces in urban areas, mostly by outlining distance thresholds (Rad and Alimohammadi 2021; Toftager et al. 2011) as well as provision (Braquinho et al. 2015; Heckert 2013) and quality standards (Engelberg et al. 2016; Nguyen et al. 2021).

2.1.2 Procedural Justice

Procedural justice reflects the notion of fairness, consistency, and representativeness in the decision-making processes that resolve disputes or allocate resources. It accentuates the importance of recognizing and respecting the differences in perspectives and beliefs across different social groups and providing affected communities with the opportunity to raise their concerns during the decision-making processes. Procedural justice aims to ensure that all stakeholders can access all the essential information and participate in the decision-making processes (Preston 2015; Schlosberg 2004).

2.1.3 Recognition Justice

Recognition justice alludes to the meaningful inclusion of different stakeholders' voices and opinions in the decision-making processes. Procedural justice and recognition justice seem to be quite similar. However, the key difference between the two dimensions is that procedural justice is more concerned with the decision-making processes, while recognition justice focuses particularly on the interests and needs of the people (Honneth 2016; Schweiger 2019).

3 Materials and Methods

This systematic review followed the Preferred Reporting Items for Systematic Reviews and Meta-Analyses (PRISMA) guidelines. It focused on articles researching the equities in green space exposure and access in the Global South.

3.1 Search Strategy

From three key terms "equity", "green space", and "global south", the researcher developed the below search query using the Boolean logic. The keywords were selected based on previous systematic reviews exploring green space-related issues to ensure validity.

The search engines used to identify relevant articles include Scopus and Web of Science since they contain peer-reviewed articles in many different disciplines that have extensively explored green-space related issues, including environmental sciences, urban planning, social sciences, and so on. The search was conducted on April 4, 2022 and updated on November 1, 2022 to obtain relevant articles published during the screening and analysis periods.

3.2 Screening Guidelines

This research used the PRISMA guidelines to select and screen papers (Fig. 2). Selected titles and abstracts were screened to ensure that they met the specific inclusion criteria, including green space, equities, and the origin of case studies. The subsequent Sect. 3.3 provides detailed information on the inclusion criteria. Titles were required to fit into at least one of the three key criteria to be eligible for the abstract screening. Abstracts that met all three key criteria were selected for the full-text assessment. Screening and selecting articles this way optimize the coverage of the search and ensure that all potentially relevant articles were captured.

3.3 Inclusion and Exclusion Criteria

To assure consistent appraisal and deliver a reproducible method, this research defined specific inclusion and exclusion criteria. Different criteria were considered during the title, abstract, and full-text screening stages (Table 1). In addition, this study only included and evaluated peer-reviewed articles to guarantee the robustness and quality of the research. Additionally, articles should be published from 2012 to 2022. This time frame is suitable since it reflects when research on this area began to gain momentum. Considerable advances in studies on inequities in green space access have happened in the past decade as well. Moreover, this research only included articles written in English. This is one of the limitations of the research, as relevant articles on this topic might have been published in the language of the studied Global South country. However, it was not practical to identify, translate, and accurately assess articles written in different languages. Therefore, this study only considered articles published in English.

Table 1 Search query

Search query	("urban greening" OR "urban green space" OR "urban greenspace" OR "urban green infrastructure" OR "park" OR "garden" OR "nature-based solution" OR "green wall" OR "green roof" OR "green") AND ("peace" OR "justice" OR "inequity" OR "equity" OR "inequities" OR "equities") AND ("global south") OR ("Developing countr*" OR vietnam* OR chin* OR colombia* OR india* OR bangladesh* OR iran* OR indonesia* OR philippin* OR mexic* OR brazil* OR chile* OR peru* OR ecuador* OR argentin* OR peru* OR cuba* OR guatemala* OR venezuela* OR nigeria*)

3.4 Data Extraction

Articles were coded based on different categories relating to descriptive statistics, study design, and outcomes of the included article. The specific codes are listed in Table 2.

To address the research objectives, this systematic review recorded the patterns of equity (inequity, equity, or mixed/no significant findings) in the included studies for each of the factors recognized in the included studies as affecting green space access (e.g., socioeconomic backgrounds, social factors, physical factors). It then synthesized the number of counts for each environmental justice dimension and explored the specific measures used to determine equities. Since environmental justice is a broad construct and there are many different methods to measure each dimension, this research pre-defined three main environmental justice dimensions and devised sub-categories to describe the open-ended description of measures. To ensure that the sub-categories reflect the answers accurately, this research reviewed classifications used in previous systematic reviews, concepts of environmental justice, and methods of the studies. This research also sought to identify the main justice issues urban green spaces in cities of the Global South were associated with.

4 Results

4.1 Descriptive Statistics

Overall, the search of the two databases resulted in a total of 3702 articles (Fig. 1). After duplicates were removed, there were 2452 articles left for screening. Reviewing the articles at the title and abstract stages decreased the number of articles eligible for the full-text screening to 121. One of the most common reasons for excluding articles was that the articles did not explain the environmental justice dimensions of urban greening. Another reason was that the articles did not list the determinants of inequities. Details about the articles included in this systematic review can be found in the annex.

Table 2 Inclusion and exclusion criteria applied at each screening stage

Key element	Inclusion criteria	Exclusion criteria
Title and abstract stage (title needs to meet at least one of the three criteria, while abstract needs to meet all the criteria)		
Green space	Presence of green space-related terms (e.g., greenspace, garden, park, green roof, green wall, wetland, etc.)	No green space-related terms
Equity	Presence of equity-related terms (e.g., disparity, inequality, equality, inequity, fair, justice, etc.)	No equity-related terms
Global South	Case study is in the Global South	Case study is not in the Global South
Full-text stage (full-text article needs to meet all the listed criteria in the title stage)		
Green space	• Exploration of at least one urban green space type (e.g., park, garden, schoolyard, and street greenery)	No mentions of green space-related terms
Equity	• Same as the title and abstract screening stage • Empirically exploration of environmental justice outcomes • Mentions of predisposing determinants of equity in urban greening such as demographic backgrounds, socioeconomic statuses, beliefs, etc.	• Same as the title and abstract screening stage • No investigation of environmental justice outcomes • No exploration of factors affecting equities in urban greening
Global South	• Use one or more case studies in the Global South to produce an in-depth understanding of the topic	• No case studies in the Global South are presented • Articles that only briefly mention cases in the Global South without applying certain methods or analysis techniques to the case studies
Study design		
Research methods	Primary studies using a quantitative, qualitative, or mixed-methods design are considered eligible. However, there must be at least a case study from the Global South. To ensure the quality of the research, only peer-reviewed articles are included	Secondary studies that only synthesize evidence or data or present an opinion (e.g., literature reviews, commentaries, letters, editorials) are ineligible
Study populations	All populations, regardless of their socioeconomic and demographic background. This research also accepts studies that do not clarify sample sizes, as long as they are relevant to the studied topic	None
Date	Articles from 2012 to 2022	Articles not published from 2012 to 2022
Language	Articles in English	Articles not in English

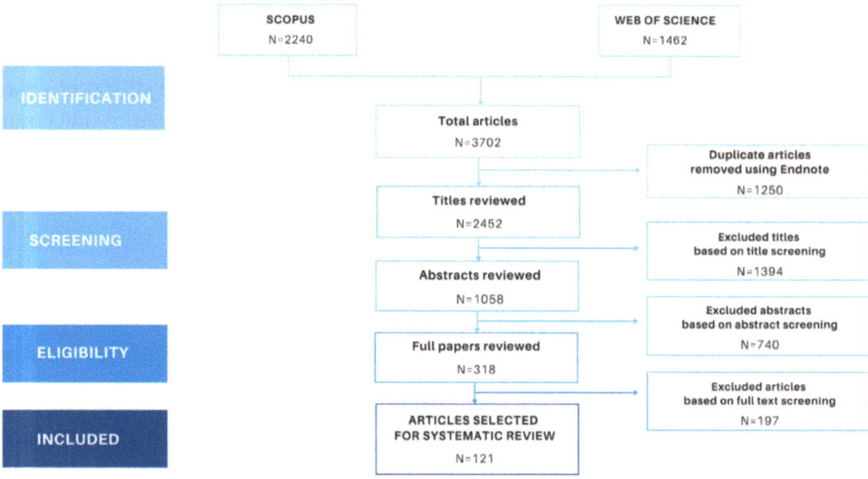

Fig. 1 PRISMA flow chart elucidating the processes of searching, including, and excluding articles

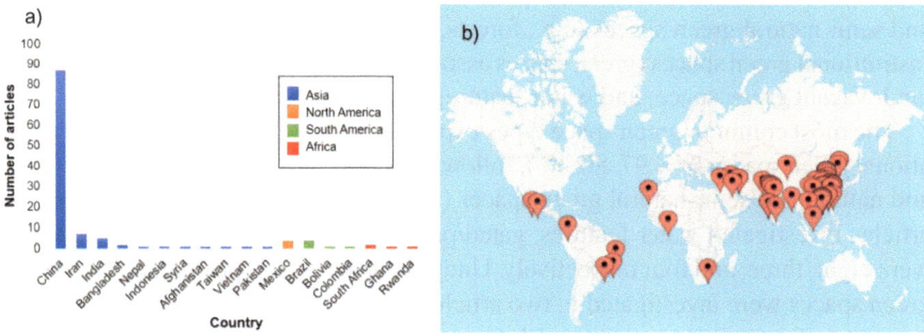

Fig. 2 Geographical locations of the 121 included studies in **a** graph; **b** world map

4.2 Geographic Distribution

Concerning the geographical location of the case studies, the included studies were carried out in 18 countries spanning Asia, North America, South America, and Africa. 108 articles (89.3%) were conducted in Asia, with most of them (n = 87, 71.9%) coming from China. Seven articles (5.8%) focused on Iran, five (4.1%) looked at case studies in India, and two articles (1.7%) investigated cases in Bangladesh. Other Asian countries were explored in one article (0.8%) each. There were four studies (3.3%) looking at a North American country, which was Mexico. Six articles (5%) delved into case studies in South America, and four of them (3.3%) looked at Brazil. Bolivia and Colombia were investigated in one article each. There were four case studies (3.3%) from Africa, with two of them (1.7%) from South Africa. One article (0.8%) looked at Ghana, and one (0.8%) focused on Rwanda. 120 articles (99.2%)

investigated one country only, whilst one article (0.8%) was comparative research between two countries (Indonesia and India). Out of the 121 articles, there were nine cross-national studies (7.4%), meaning that they investigated different cities within a country. The geographical allocation was highly skewed toward Asian countries, especially China. This could be due to the large population size of China and the absence of data available in other countries. Figure 2 demonstrates the specific geographic distribution of the included studies.

4.3 Types of Green Spaces

Different studies used different typologies of green spaces. To simplify the process of managing and analyzing records, this research developed a classification system based on the functions and characteristics of green spaces. There are eight types of green spaces: recreational green spaces (e.g., parks, gardens); open green spaces (e.g., street trees, green corridors, green verges, green roofs); protected areas (e.g., national parks dedicated for conservation purposes and protected by the governments); natural and semi-natural green spaces (e.g., forests, woodland); gated/private green spaces; institutional green spaces (green spaces associated with a school or business); unused land/ vacant green space; and water features (e.g., wetland).

The most common green space types explored in the included studies were recreational green spaces (n = 97, 80.2%), followed by open green spaces (n = 25, 20.7%) and natural and semi-natural green spaces (n = 11, 9.1%) (Fig. 3a). The number of articles investigated water features, gated/private green spaces, and protected areas were eight, three, and four respectively. Unused/vacant green spaces and institutional green spaces were investigated in two articles each. There was an unevenness in the coverage of green space types, with most articles focusing on recreational green spaces such as parks and gardens. 22 studies (18.2%) looked at multiple types of green spaces. Figure 3b shows the correlation between frequency of green space types and that of different environmental justice aspects. Within the included articles, recreational green spaces were the type of green space most often investigated in articles focusing on justice in distribution (n = 95, 78.5%), procedure (n = 5, 4.1%), and recognition (n = 13, 10.7%). There were only three articles looking at protected areas, but within them, issues related to distributional, procedural, and recognition justice were discussed quite evenly.

4.4 Study Design

Most of the studies (n = 110, 90.2%) were quantitative, with four using longitudinal data. Six studies (5%) adopted a qualitative approach, and five studies (4.2%) were mixed-methods. Quantitative studies used secondary data sources such as census data (n = 102, 84.3%) and mobile phone data (n = 5, 4.1%). There were 10 quantitative

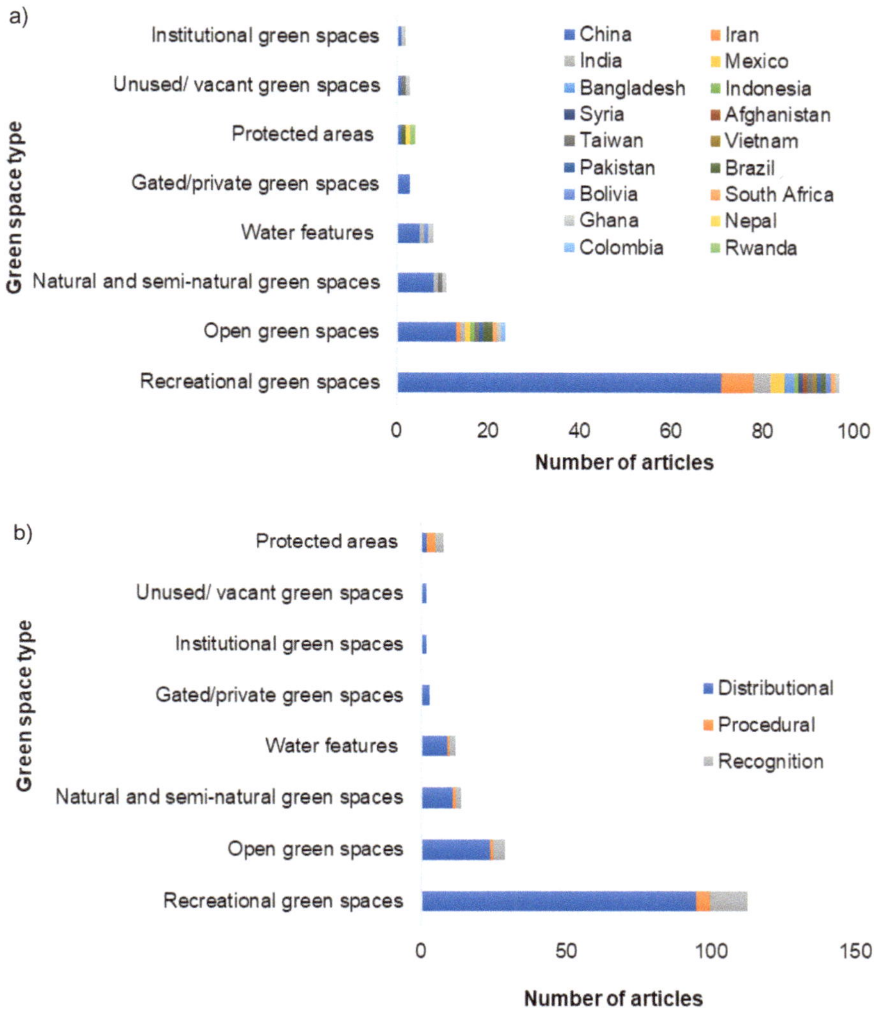

Fig. 3 Frequency of green space type in the included 121 studies by **a** country, **b** environmental justice dimension

studies (8.3%) using primary data from surveys. Qualitative studies used primary data sources like interviews (n = 6, 5%) and surveys (n = 2, 1.7%). Mixed-methods studies combined the use of surveys, interviews, and census data.

There are four main types of data analysis techniques. Quantitative studies either used spatial (n = 23, 19%) or statistical analysis (n = 19, 15.7%). There were 70 studies (57.9%) that used both techniques. All qualitative studies (n = 6, 5%) relied on content analysis, while mixed-methods studies (n = 5, 4.2%) combined statistical analysis with content analysis to interpret the data.

4.5 Environmental Justice Dimensions

Most articles (n = 116, 95.9%) investigated equities in urban greening from a distributional justice perspective. Eight articles (6.6%) adopted a procedural lens, and 18 articles (14.9%) adopted a recognition justice dimension. 19 articles (15.7%) assessed more than one environmental justice dimension. There are a variety of methods which the authors used to evaluate different environmental justice dimensions (Table 3).

4.6 Distributional Justice

The indicators of distributional justice include proximity, quality, quantity, and perceived accessibility. Out of them, proximity is the indicator investigated the most. Notably, most methods were quantitative. Particularly, to measure proximity, quantity, perceived accessibility, all authors relied on quantitative approaches using statistical and spatial analysis.

Green space qualities were measured using quantitative, qualitative, or both approaches. In terms of green space qualities, most authors focused on size (n = 12), ecosystem services provided by green spaces (n = 11), presence of amenities (n = 9), and vegetation coverage (n = 5). 17 articles measured the quantity of green spaces using quantitative methods. There was only one article looking at perceived accessibility.

4.7 Procedural Justice

Out of the three environmental justice dimensions, procedural justice was the one investigated the least with only eight articles (6.6%). They addressed this environmental justice dimension by measuring the degree to which people are willing to participate in the decision-making processes.

4.8 Recognition Justice

Regarding recognition justice, one article looked at people's satisfaction with the existing green spaces. The remaining articles explored how the existing green spaces met their needs. One article scrutinized both dimensions.

Table 3 Codes used for analysis of the included studies

Category	Detail	Coding options
Descriptive codes	Name of author, name of article, year of publication, etc.	Free text description
Study design	Data collection approach of the study	Three options (quantitative, qualitative, or mixed-methods)
Data sources	Data can come from primary sources (interviews, surveys, etc.) or secondary sources (census data, remote sensing, etc.)	• Two options (primary and secondary) • Open-ended description of data sources (interview, survey, geospatial data, etc.)
Data analysis tools	Analysis techniques used to interpret the data	Open-ended description of tools used to analyze the data (e.g., statistical analysis, spatial analysis)
Case study information	Location of case study, type of green space, population, etc.	Free text description
Outcomes on green space access	Inequity (1: disadvantaged populations have less access; –1: disadvantaged populations have more access), or mixed findings/not significant (0)	Three options
Aspects of environmental justice explored in the articles and indicators of each aspect	Distribution, procedural, and recognition justice	• Three options for environmental justice aspects (distribution, procedural, and recognition) • Open-ended description of the indicators used to quantify the environmental justice aspect(s)

(continued)

Table 3 (continued)

Category	Detail	Coding options
Predisposing factors	Predisposing factors are grouped into three main categories: personal (gender, age, economic status, ethnic minority, education, immigration status), physical (transport network, travel mode), and social factors (population density, intra-urbanization)	• 10 factors as described in the detail section • Open-ended description of the factors
Environmental justice issues	From the initial screening of full-text articles, six categories of environmental justice issues were found: health and wellbeing; accessibility (degree of accessibility for green space users to enjoy quality and distance); food; economic benefits; social benefits such as aesthetics cultivation, social interaction, and recreation; environmental benefits such as improved air quality, water quality, etc.	• Six categories as described in the detail section • Open-ended description of the specific issues the article addresses

4.9 *Predisposing Factors and Their Association with Equities in Urban Greening*

Predisposing factors refer to the characteristics or conditions of a person, society, or physical infrastructure that either promote or hinder the use and/or access of green space. This study identified three main types of predisposing factors: personal, social, and physical. Personal factors entail gender, age, economic status defined in terms of income, housing ownership, or unemployment rate, education level, and immigration status. Social factors are external influences related to social structure and processes. This review recognized two social factors: new development and population density. Physical factors refer to tangible infrastructure. This review found one physical factor, which is transport network. Table 4 summarizes the predisposing factors found in the 121 included articles, their respective environmental justice dimensions, and their correlation between green space access.

Table 4 Different methods used to assess environmental justice dimensions

Environmental justice dimension	Environmental justice sub-category	Type of	Methods	Number of articles mentioning
Distributional justice (n = 116, 95.9%)	Proximity	Quantitative	Travel time, travel cost	13
			Distance to the nearest green space using different transport modes	23
			Euclidean distance	1
			Two-step floating catchment area	26
			Buffer analysis	9
			Network analysis	10
			Distribution by population size	22
	Quality	Quantitative (7)/ qualitative (1)	Presence of amenities (e.g., playgrounds, natural features)	8
		Quantitative	Level of maintenance	1
		Quantitative/ qualitative	Safety	1
		Quantitative	Vegetation coverage or naturalness of green space (normalized difference vegetation Index)	10
		Quantitative	Presence of vegetation (green view index)	5
		Quantitative	Size (per capita green space/ coverage ratio of park area)	12
		Quantitative/ qualitative	Ecosystem services provided by green space (recreational opportunity, green space service coverage)	11
	Quantity	Quantitative	Acreage of green space	17

(continued)

Table 4 (continued)

Environmental justice dimension	Environmental justice sub-category	Type of	Methods	Number of articles mentioning
	Perceived accessibility	Quantitative	People's subjective perceptions	1
Procedural justice (n = 8, 6.6%)	Inclusive participation in the decision-making processes	Qualitative	Willingness to participate in the decision-making processes/the level of community engagement	8
Recognition justice (n = 18, 14.9%)	Demonstrated needs and satisfaction	Qualitative	Satisfaction with green spaces	1
		Qualitative	Preferences for certain features of green spaces (e.g., the need for more amenities, natural features, or safety features)	18

4.10 Personal Factors

Out of all personal factors, most studies (n = 75, 62%) investigated economic statuses. This is followed by age (n = 26, 21.5%), education (n = 18, 14.9%), transport access (n = 17, 14%), immigration status (n = 12, 9.9%), and gender (n = 10, 8.3%). Only nine articles (7.4%) looking at racial or ethnic minority, which is a big contrast compared to studies looking at countries in the Global North. All articles indicated that there are mixed patterns in the association between green space access and all personal factors. This research also found that most articles were concerned with the distributional dimension of justice.

4.11 Social Factors

A total of 22 studies (18.2%) investigated population density, while six studies (5%) explored new development. 19 articles focusing on population density found that people living in areas of a high population density have lower access to green spaces, while three articles found no differences. On the other hand, all articles looking at new development found that there were disparities in access to green spaces between old and new neighborhoods. Most articles analyzed the issues from a distributional justice perspective, and only one article investigating the recognition justice dimension.

4.12 Physical Factor

All five articles looking at transport network focused on distributional justice and demonstrated that the lack of transport network hindered people from accessing green spaces.

4.13 Sustainability Issues Discussed in the Included Articles

Urban green spaces have been associated with many benefits, spanning from improving the environmental quality to supporting the health and wellbeing of urban citizens. Table 5 indicates the specific benefits of urban green spaces discussed in the studied articles. Since they addressed a wide variety of contributions of green space to sustainability, this research categorized them into six main groups, including spatial inclusion, environmental quality, health and wellbeing, social benefits, food provision, and economic benefits. The green space benefit most commonly addressed in the studied articles were spatial inclusion (n = 97, 80.2%), followed by environmental quality (n = 13, 10.7%), health and wellbeing (n = 12, 9.9%), and social benefits (n = 10, 8.3%). Two articles were concerned about food provision, while one article mentioned economic benefits of green spaces.

Within the studied articles, distributional justice issues were addressed most often regarding spatial inclusion (n = 95, 78.5%) and environmental quality (n = 12, 9.9%) (Table 6). For procedural justice, spatial inclusion (n = 8, 6.6%) and health and wellbeing (n = 3, 2.5%) were the most commonly mentioned issues in articles looking at procedural justice. Regarding articles focusing on recognition justice, the most frequently addressed type of issue was social benefits (n = 7, 5.8%). This was followed by health and wellbeing (n = 6, 5%) and environmental quality (n = 5, 4.1%) (Table 6).

5 Discussion

5.1 Environmental Justice Dimensions in Cities of the Global South

This research shows many studies exploring equities in urban greening in cities of the Global South have been concerned with the geographic distribution of green spaces across different regions and across different populations. In terms of sustainability issues green spaces can help to address, most of the included articles focused mainly on spatial inclusion although the recognition of green space benefits in terms of improving people's health as well as providing social and environmental benefits have begun to emerge. In general, in cities of the Global South, the concept of "green

Table 5 Predisposing factors, environmental justice dimensions, and association between predisposing factors and access to green spaces

Predisposing factors		Number of articles mentioning	Environmental justice dimensions (number of articles mentioning)	Patterns of equity in green space access (number of articles finding)	Detail
Personal factors	Gender	10	Distributional (9), recognition (5)	Inequitable (9)	Women have less access to green space compared to men
				Not significant (1)	There are no differences in green space access between men and women
	Age (young people and children under 18, the elderly above 60)	26	Distributional (26), recognition (4)	Inequitable (22)	Children/the elderly have less access to green spaces compared to other age groups (21)
					Children/the elderly have more access to green spaces compared to other age groups (1)
				Not significant (4)	There are no differences in green space access across different age groups
	Economic status (income, housing ownership, unemployment rate)	75	Distributional (73), procedural (5), recognition (10)	Inequitable (65)	Those of low economic statuses have less access to green spaces (60)

(continued)

Table 5 (continued)

Predisposing factors		Number of articles mentioning	Environmental justice dimensions (number of articles mentioning)	Patterns of equity in green space access (number of articles finding)	Detail
					Those of high economic statuses have less access to green spaces (5)
				Not significant (8)	There are no differences in green space access across people of different economic statuses
	Racial or ethnic minority	9	Distributional (8), recognition (2)	Inequitable (8)	Those of racial/ethnic minority background have less green space access (6)
					Those of racial/ethnic minority background have more green space access (2)
	Education	18	Distributional (17), procedural (2), recognition (7)	Inequitable (15)	Those of low educational levels have less green space access (14)
					Those of low educational levels have more green space access (1)
				Not significant (2)	There are no differences in green space access
	Immigration status	12	Distributional (10), procedural (4), recognition (5)	Inequitable (8)	Immigrants have less access to green spaces

(continued)

Table 5 (continued)

Predisposing factors		Number of articles mentioning	Environmental justice dimensions (number of articles mentioning)	Patterns of equity in green space access (number of articles finding)	Detail
				Not significant (2)	There are no differences in green space access
	Transport access	17	Distributional (17), recognition (2)	Inequitable (13)	Those driving cars tend to have more access to green spaces compared to pedestrians and cyclists
				Not significant (4)	There are no differences in green space access between different transport modes
Social factors	New development	6	Distributional (6), recognition (1)	Inequitable (6)	New developments attract more investment, leaving old neighborhoods with decayed green spaces
	Population density	22	Distributional (22)	Inequitable (19)	Those living in areas of high population density have less access to green spaces

(continued)

Table 5 (continued)

Predisposing factors		Number of articles mentioning	Environmental justice dimensions (number of articles mentioning)	Patterns of equity in green space access (number of articles finding)	Detail
				Not significant (3)	There are no differences between areas of different population density
Physical factor	Transport network	5	Distributional (5)	Inequitable (5)	The absence of adequate transport networks hinder access to green spaces

equity" just refers to the equitable allocation of green spaces and has not expanded to encompass equitable distribution of green space benefits. It is important to raise awareness about other green space functions and the social equity issues associated with them, as the need for measures to mitigate risks of natural disasters and climate change as well as to enhance the livability and resilience of urban areas becomes more prominent.

Another important finding is that within the studied articles, procedural justice and recognition justice were addressed in a much more limited extent for all issues concerning green spaces. Importantly, issues concerning procedural justice and recognition justice have started to become a focal point in the field of urban planning and are discussed extensively in literature focusing on cities in the Global North. It is important to recognize that these environmental justice dimensions are crucial in the context of cities in the Global South since many of them are non-democratic and there are uneven power relations between decision makers and the public. The government usually devises the country's priorities and plans developments without considering the needs of different social groups. Planning for an inclusive and just city requires active participation of urban citizens (Lake 2016) and equitable distribution of benefits from green spaces (Rodenbiker 2022). As a result, the development of green spaces in cities of the Global South should aim to provide environmental benefits and foster openness and inclusivity. This can only be achieved through facilitating inclusion and public participation in the decision-making processes.

Also, this study shows that literature of cities in the Global South mostly focuses on social groups such as the low-income people, the elderly, children, and people of different education levels. Meanwhile, there is a lack of studies investigating the perspectives of women or those living in areas of a high population density. No

Table 6 Breadth of green space benefits discussed in the included 121 articles

Sustainability issues	Definition	Number of articles	Environmental justice dimension		
			Distributional	Procedural	Recognition
Spatial inclusion	Providing equitable access to resources for different populations	97	95	8	3
Environmental quality	Providing environmental benefits such as air purification, noise reduction, improvement in water quality, runoff mitigation, temperature regulation, flood regulation, carbon sequestration, and dust filtering	13	12	2	5
Health and wellbeing	Supporting people's mental and physical health	12	11	3	6
Social benefits	Providing the local area with benefits such as cultivation of aesthetics, opportunities for social interaction, etc.	10	10	1	7
Food provision	Providing opportunities for food growing	2	2	1	1
Economic benefits	Increasing land prices to support the local economy	1	1	0	0

studies in the Global South have explored groups such as people of different cultural backgrounds or disabled people. Recognizing the voices and perspectives only of those interested in ensuring equities in green space access might compromise the potential of urban greening practices in cities of the Global South in addressing the current equity challenges and promoting use. Besides experts and current green space users, it is important to seek out perspectives from those who are not using green spaces and are underrepresented in the decision-making processes to ensure that green spaces can address diverse needs and provide various benefits for the local environment.

5.2 Principles of Justice-Focused Urban Greening Practices: The Role of Decision-Makers and Urban Developers

The questions of who benefits from urban green spaces, who regulates and decides urban greening planning practices, and whose needs are recognized in the decision-making processes are core to an environmental justice agenda. As indicated in many studies, the current planning practices in the Global South are project-bounded and driven by experts and decision-makers. Such universal green space planning practices might alienate certain social groups and perpetuate the structural disparities that exist within the urban society. It is, therefore, important to reorient planning practices and establishing justice as one of the core principles. In fact, justice-oriented approaches to developing and distributing green spaces can help reduce socioeconomic inequities due to urbanization by ensuring that all voices are heard (Rodenbiker 2022). Based on the findings of this research, a conceptual framework outlining principles necessary to develop inclusive and sustainable green spaces were developed (Fig. 4).

This framework links the concept of sustainable development and environmental justice together and advocates for the inclusion of all three environmental justice dimensions. Currently, environmental justice is not integrated within the language of the Sustainable Development Goals (SDGs). There are also fundamental differences regarding how the two concepts were developed. While sustainable development is a top-down, global policy framework, environmental justice is a grassroots reaction to environmental apartheid (Menton et al. 2020). Both concepts are also associated with different goals. Environmental justice focuses particularly on equality and highlights that everyone, regardless of their backgrounds, should have equal access to environmental resources and benefits (Schlosberg 2004). Meanwhile, sustainable

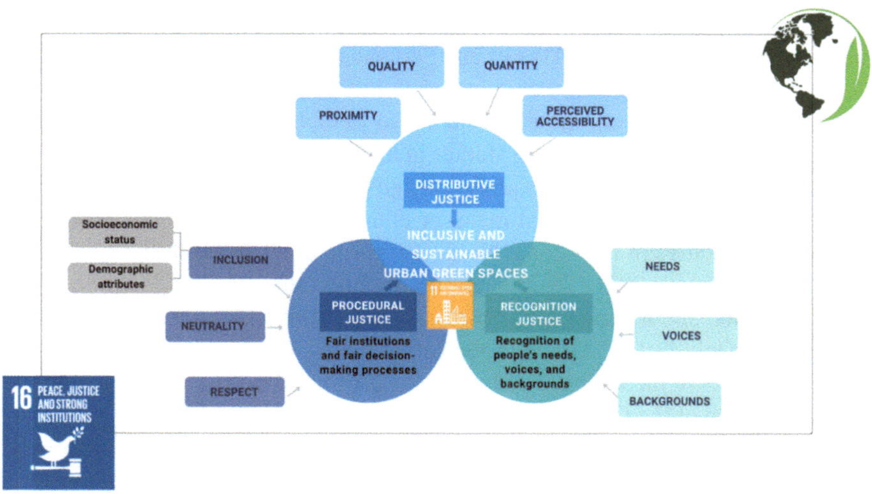

Fig. 4 A conceptual framework of principles for inclusive and sustainable green spaces

development encompasses three overarching pillars: economic, social, and environment and attempts to improve people's life and wellbeing without depleting natural resources or damaging the environment (United Nations 2015).

Nonetheless, there is a common ground between sustainable development and environmental justice, particularly regarding distribution of resources, justice, and the role of institutions and communities. Undeniably, justice is an underpinning principle behind all the SDGs. Particularly, goal 16 clearly articulates the cruciality of developing "inclusive, participatory, and representative decision making" and of ensuring "effective, accountable, and transparent, institutions at all levels" (United Nations 2015). Justice is the fundamental yet quintessential component of the achievement of sustainable development and underpins the adjudication of social, economic, and environmental equity or inequity (Schlosberg 2004). Overall, both sustainable development and environmental justice are symbiotically linked to each other and can work in tandem to achieve a synergistic impact (Fisher 2003).

To realize goal 11 "sustainable cities and communities", it is crucial to consider diverse perspectives in the decision-making processes to enhance social inclusion and ensure the equitable distribution of greenspace benefits. And to facilitate such process, the institutions need to be transparent and consider all respect people of different backgrounds. Undeniably, involving citizens in the decision-making process can compound the benefits of green space while reducing the inherent social disparities and leveraging social capacity.

5.3 *Comparison between Global South and Global North Cities*

Similar to previous studies on social disparities in green space access in countries of the Global North (Fernández-Álvarez and Fernández-Álvarez 2017; Kim et al. 2021; Mears et al. 2019), studies investigating cases in the Global South focus mostly on the distributional justice dimension, with a heavy emphasis on proximity.

Regarding predisposing factors, studies in the Global South focused on people of different socio-economic statuses, different age groups, and education levels. There were not many studies looking at racial or ethnic backgrounds, which is different from studies in cities of the Global North (Ferguson et al. 2018; Miller et al. 2022; Rigolon and Németh 2021). This might be because cities of the Global South are more concerned about economic growth and populations (Arku and Marais 2021; Yazdani and Dola 2013). Although racism is a persistent structural and systemic problem in cities of the Global North (Winant 2006), it is not as blatant and fatal of an issue in cities of the Global South.

Many studies have demonstrated that in the Global North, there is a mixed pattern of equity for green space proximity (Mears et al. 2019; Rigolon 2016). But regarding green space qualities, it is observed that the disadvantaged communities are always exposed to green spaces of low qualities (Engelberg et al. 2016; Nguyen et al. 2021;

Rigolon 2016). Interestingly, this research shows that in countries of the Global South, there are mixed findings for both green space proximity and qualities. Although most studies still indicated inequitable outcomes, some argued that access to green spaces, in terms of both proximity and qualities, do not vary significantly between different social groups. The differences in spatial patterns of landscape and urban development trends might be the reasons behind these different results (Chen et al. 2022; Eva et al. 2022; Yazdani and Dola 2013). Different methods used to measure green space qualities might be another reason.

5.4 *Strengths, Limitations, and Future Scope*

The strength of this research is that it comprehensively synthesized evidence and addressed many issues concerning environmental justice in urban greening of cities in the Global South. First, it uncovered the different environmental justice dimensions of urban greening in the Global South. Second, it analyzed the relationships between environmental justice dimensions and factors recognized as influencing green space disparities. What further set this study apart was that it looked at not only personal factors such as socioeconomic statuses and racial or ethnic background, but also social or physical factors that were not conciliated by individual-level attributes. Social and physical factors including the history, urban developmental trend, and infrastructure of a city play a key part in influencing its urban planning agenda. Therefore, understanding those factors can help inform urban planning and improve the current practices. This research also proposed strategies to improve urban greening practices in cities of the Global South.

Nonetheless, this research has some limitations that future research could help resolve. First, the decision to limit the searches to articles published in English might introduce a bias by potentially excluding relevant articles written in another language spoken in Global South countries such as Vietnamese, Chinese, and so on. Most of the included articles also focused on cities in China, so the findings might speak mostly to those cities. Hence, future research could investigate articles published in languages other than English to provide a more comprehensive review of this topic and verify the findings of this study. Moreover, a meta-analysis was not conducted due to the variability in green space typologies and methods across the included studies. Future research could attempt to reframe the question and redefine the inclusion and exclusion criteria to include only papers with a consistent population across intervention and outcomes. This way, a formal meta-analysis could be achieved.

This paper also emphasizes the needs for more longitudinal studies, as the majority of the included studies were cross-sectional and did not track the changes in urban greening practices overtime. Additionally, most studies adopted a quantitative approach to measure the spatial distribution of green spaces, and there were only a few using qualitative methods or mixed-method approaches to explore people's perspectives and willingness to participate in the decision-making processes. Understanding the features of green spaces that are appealing or alienating to people and knowing their thoughts on green spaces are crucial for community-based green space planning. While quantitative methods and existing data can provide a numerical evidence base to inform planning, they cannot account for the wide array of people's behaviors, perspectives, and experiences and cannot provide information on what could be done to alleviate these in different contexts.

Many cities in the Global South have been facing great socio-economic and environmental problems, including increase in slums, increasing urban poverty, lack of natural resources, air and water pollution, and urban marginalization. How to make Global South cities become more functionally integrated, socially inclusive, as well as environmentally and economically sustainable is a pressing issue for decision makers and urban developers. Unfortunately, many governments in the Global South are fragmented and lack the capacity to address comprehensively censorious urban problems such as environmental degradation, climate change, social inequity, and poverty. While urban greening can contribute to making the cities more resilient and sustainable as established in Sustainable Development Goal 11 and the New Urban Agenda, there needs to be a shift away from the conventional planning approaches to more inclusive planning systems that involve citizens in devising solutions for pressing urban challenges (Arku and Marais 2021).

It is challenging to change the nature of urban planning practices in cities of the Global South since the prevailing planning pedagogies are guided by hegemonic directives as well as institutional legislations and requirements. Justice is an abstract notion and achieves its aims and intentions through discussions rather than being established as static facts. For justice to be formulated as a guiding planning principle, decision makers and urban planners need to actively engage in conversations with the communities and co-creating solutions instead of pre-determining ones. They should also ensure all participants are included in the processes of planning for their local environment (Lake 2016).

Qualitative methods, including social surveys, public planning discussion forums, and/ or consultation sessions targeting specifically the disadvantaged and underrepresented communities can be effective in garnering the public's opinions (Rodenbiker 2022). Creative participatory approaches like co-creation and co-design are also gaining popularity in both academia and practice as potential methods to truly involve and engage with participants. They are built on the idea of democracy and social justice and create a platform for freedom of expression (Heimburg et al. 2021). Regardless of the qualitative methods used, it is crucial to understand who participate in the research process and how they participate. Furthermore, there needs to be an impact assessment to evaluate the outcomes of involving different participants as well as how their perspectives are incorporated. Overall, participatory approaches can facilitate active citizenship, increase social justice and public values if they are properly implemented and assessed.

6 Conclusion

Not many studies on urban greening and environmental justice have focused on cities of the Global South. This research has filled in the gap by synthesizing the existing evidence on the topic of social equities in urban greening and appraising it against the different environmental justice dimensions, which have been discussed extensively in studies of the Global North but not so much in the case of the Global South. It highlights that the distributional justice was the environmental justice dimension most investigated in the literature, while there is little attention paid to procedural and recognition justice. However, it is important to spread the focus evenly across all dimensions of environmental justice, particularly because procedural and recognition justice can help mold the way people experience distributional justice and build the foundation for distributional justice to be realized.

Many societies, both in the Global South and Global North, have not been able to resolve the widening and tenacious issue regarding social equity in access to green spaces. As stipulated in SDG 16, justice is a key factor for sustainable development. There needs to be an improvement in institutions' capacity and a strong focus on collaboration to advance fairness, social justice, and sustainability. This paper encourages innovation, public participation, and public values to be incorporated in urban planning.

References

Akpinar A (2016) How is quality of urban green spaces associated with physical activity and health? Urban Forest Urban Green 16:76–83. https://doi.org/10.1016/J.UFUG.2016.01.011

Antuna-Rozado C, Herzog CP, Freitas T, Cagnin C, Wiedman G (2019) Nature based solutions (NBS) for sustainable and resilient cities: experiences from Europe and Brazil. IOP Conf Ser: Earth Environ Sci 297(1). https://doi.org/10.1088/1755-1315/297/1/012001

Arku G, Marais L (2021) Global south urbanisms and urban sustainability—Challenges and the way forward. Front Sustain Cities 3. https://doi.org/10.3389/frsc.2021.692799

Artmann M, Mueller C, Goetzlich L, Hof A (2019) Supply and demand concerning urban green spaces for recreation by elderlies living in care facilities: the role of accessibility in an explorative case study in Austria. Front Environ Sci 7:136. https://doi.org/10.3389/FENVS.2019.00136/BIBTEX

Braquinho C, Cvejić R, Eler K, Gonzales P, Haase D, Hansen R, Kabisch N, Lorance Rall E, Niemela J, Pauleit S, Pintar M, Lafortezza R, Santos A, Strohbach M, Vierikko K, Železnikar Š (2015) A typology of urban green spaces, eco-system services provisioning services and demands

Byrne J, Wolch J (2009) Nature, race, and parks: past research and future directions for geographic research 33(6):743–765. https://doi.org/10.1177/0309132509103156

Chen B, Wu S, Song Y, Webster C, Xu B, Gong P (2022) Contrasting inequality in human exposure to greenspace between cities of Global North and Global South. Nat Commun 13(1):4636. https://doi.org/10.1038/s41467-022-32258-4

Cole HVS, Garcia Lamarca M, Connolly JJT, Anguelovski I (2017) Are green cities healthy and equitable? Unpacking the relationship between health, green space and gentrification. J Epidemiol Commun Health, jech-2017-209201. https://doi.org/10.1136/jech-2017-209201

Curran W, Hamilton T (2012) Just green enough: contesting environmental gentrification in Greenpoint, Brooklyn. Local Environ 17(9):1027–1042. https://doi.org/10.1080/13549839.2012.729569

Dadvand P, Wright J, Martinez D, Basagaña X, McEachan RRC, Cirach M, Gidlow CJ, de Hoogh K, Gražulevičiene R, Nieuwenhuijsen MJ (2014) Inequality, green spaces, and pregnant women: roles of ethnicity and individual and neighbourhood socioeconomic status. Environ Int 71:101–108. https://doi.org/10.1016/J.ENVINT.2014.06.010

de Vries S, Buijs AE, Snep RPH (2020) Environmental justice in The Netherlands: presence and quality of greenspace differ by socioeconomic status of neighbourhoods. Sustainability (Switzerland) 12(15). https://doi.org/10.3390/SU12155889

Engelberg JK, Conway TL, Geremia C, Cain KL, Saelens BE, Glanz K, Frank LD, Sallis JF (2016) Socioeconomic and race/ethnic disparities in observed park quality. BMC Publ Health 16(1). https://doi.org/10.1186/s12889-016-3055-4

Eva M, Cehan A, Corodescu-Roşca E, Bourdin S (2022) Spatial patterns of regional inequalities: empirical evidence from a large panel of countries. Appl Geogr 140:102638. https://doi.org/10.1016/j.apgeog.2022.102638

Feng S, Chen L, Sun R, Feng Z, Li J, Khan MS, Jing Y (2019) The distribution and accessibility of urban parks in Beijing, China: implications of social equity. Int J Environ Res Publ Health 16(24). https://doi.org/10.3390/ijerph16244894

Ferguson M, Roberts HE, McEachan RRC, Dallimer M (2018) Contrasting distributions of urban green infrastructure across social and ethno-racial groups. Landsc Urban Plan 175:136–148. https://doi.org/10.1016/J.LANDURBPLAN.2018.03.020

Fernández-Álvarez R, Fernández-Álvarez R (2017) Inequitable distribution of green public space in Mexico City: an environmental injustice case. Econ Soc Territ 17(54):399–428. https://doi.org/10.22136/EST002017697

Fisher E (2003) Sustainable development and environmental justice: same planet, different world? Environs 26(2):201–2017

Gong F, Zheng Z-C, Ng E (2016) Modeling elderly accessibility to urban green space in high density cities: a case study of Hong Kong. Procedia Environ Sci 36:90–97. https://doi.org/10.1016/J.PROENV.2016.09.018

Heckert M (2013) Access and equity in greenspace provision: a comparison of methods to assess the impacts of greening vacant land. Trans GIS 17(6):808–827. https://doi.org/10.1111/TGIS.12000

Heimburg DV, Ness O, Storch J (2021) Co-creation of public values: citizenship, social justice, and well-being. In: Processual perspectives on the co-production turn in public sector organizations edition, vol 1. https://doi.org/10.4018/978-1-7998-4975-9.ch002

Honneth A (2016) Recognition and justice 47(4):351–364. https://doi.org/10.1177/0001699304048668

Huang JH, Hipp JA, Marquet O, Alberico C, Fry D, Mazak E, Lovasi GS, Robinson WR, Floyd MF (2020) Neighborhood characteristics associated with park use and park-based physical activity among children in low-income diverse neighborhoods in New York City. Prev Med 131:105948. https://doi.org/10.1016/J.YPMED.2019.105948

Hwang YH, Nasution IK, Amonkar D, Hahs A (2020) Urban green space distribution related to land values in fast-growing megacities, Mumbai and Jakarta–Unexploited opportunities to increase access to greenery for the poor. Sustainability 12(12):4982. https://doi.org/10.3390/SU12124982

Jones AP, Brainard J, Bateman IJ, Lovett AA (2009) Equity of access to public parks in Birmingham (UK). In: Income distribution: inequalities, impacts and incentives

Kim J, Lee KJJ, Thapa B (2021) Visualizing fairness: distributional equity of urban green spaces for marginalized groups 65(5):833–851. https://doi.org/10.1080/09640568.2021.1915258

Klein W, Dove MR, Felson AJ (2021) Engaging the unengaged: understanding residents' perceptions of social access to urban public space. Urban Forest Urban Green 59:126991. https://doi.org/10.1016/J.UFUG.2021.126991

Kronenberg J, Haase A, Łaszkiewicz E, Antal A, Baravikova A, Biernacka M, Dushkova D, Filčak R, Haase D, Ignatieva M, Khmara Y, Niţă MR, Onose DA (2020) Environmental justice in the context of urban green space availability, accessibility, and attractiveness in postsocialist cities. Cities 106:102862. https://doi.org/10.1016/J.CITIES.2020.102862

Lake RW (2016) Justice as subject and object of planning. Int J Urban Reg Res 40(6):1205–1220. https://doi.org/10.1111/1468-2427.12442

Lee ACK, Jordan HC, Horsley J (2015) Value of urban green spaces in promoting healthy living and wellbeing: prospects for planning. Risk Manag Healthc Policy 8:131. https://doi.org/10.2147/RMHP.S61654

Mayor of London (2018) London environment strategy. www.london.gov.uk

Mears M, Brindley P, Maheswaran R, Jorgensen A (2019) Understanding the socioeconomic equity of publicly accessible greenspace distribution: the example of Sheffield, UK. Geoforum 103. https://doi.org/10.1016/j.geoforum.2019.04.016

Menton M, Larrea C, Latorre S, Martinez-Alier J, Peck M, Temper L, Walter M (2020) Environmental justice and the SDGs: from synergies to gaps and contradictions. Sustain Sci 15(6):1621–1636. https://doi.org/10.1007/s11625-020-00789-8

Miller EB, Seyfried AP, Pender SE, Heard K, Meindl GA (2022) Racial disparities in access to public green spaces: using geographic information systems to identify underserved populations in a small American city. Environ Justice 15(4). https://doi.org/10.1089/env.2021.0067

Nasri Roodsari E, Hoseini P (2021) An assessment of the correlation between urban green space supply and socio-economic disparities of Tehran districts—Iran. Environ Dev Sustain. https://doi.org/10.1007/s10668-021-01970-4

Nguyen PY, Astell-Burt T, Rahimi-Ardabili H, Feng X (2021) Green space quality and health: a systematic review. Int J Environ Res Publ Health 18(21). https://doi.org/10.3390/IJERPH182111028/S1

Preston BJ (2015) The effectiveness of the law in providing access to environmental justice: an introduction. Search Environ Justice 23–43. https://doi.org/10.4337/9781784719425.00008

Rad TG, Alimohammadi A (2021) Modeling relationships between the network distance and travel time dynamics for assessing equity of accessibility to urban parks. Geo-Spatial Inf Sci 24(3). https://doi.org/10.1080/10095020.2020.1858189

Rigolon A (2016) A complex landscape of inequity in access to urban parks: a literature review. In: Landscape and urban planning, vol 153. Elsevier B.V., pp 160–169. https://doi.org/10.1016/j.landurbplan.2016.05.017

Rigolon A, Browning M, Jennings V (2018a) Inequities in the quality of urban park systems: an environmental justice investigation of cities in the United States. Landsc Urban Plan 178. https://doi.org/10.1016/j.landurbplan.2018.05.026

Rigolon A, Browning M, Lee K, Shin S (2018b) Access to urban green space in cities of the global south: a systematic literature review. Urban Sci 2(3):67. https://doi.org/10.3390/urbansci2030067

Rigolon A, Németh J (2021) What shapes uneven access to urban amenities? Thick injustice and the legacy of racial discrimination in Denver's parks. J Plan Educ Res 41(3). https://doi.org/10.1177/0739456X18789251

Rodenbiker J (2022) Social justice in China's cities: urban-rural restructuring and justice-oriented planning. Trans Plan Urban Res 1(1–2):184–198. https://doi.org/10.1177/27541223221111799

Sathyakumar V, Ramsankaran RAAJ, Bardhan R (2019) Linking remotely sensed urban green space (UGS) distribution patterns and socio-economic status (SES)—A multi-scale probabilistic analysis based in Mumbai, India. GIScience and Remote Sens 56(5). https://doi.org/10.1080/15481603.2018.1549819

Schäffler A, Swilling M (2013) Valuing green infrastructure in an urban environment under pressure. Ecol Econ. https://studylib.net/doc/25265364/valuing-green-infrastructure-in-an-urban-environment-unde

Schlosberg D (2004) Reconceiving environmental justice: global movements and political theories. Environ Politics 13(3):517–540. https://doi.org/10.1080/0964401042000229025

Schweiger G (2019) Recognition, misrecognition and justice. Ethics Glob Politics 12(4):1693870. https://doi.org/10.1080/16544951.2019.1693870

Shanahan DF, Lin BB, Gaston KJ, Bush R, Fuller RA (2014) Socio-economic inequalities in access to nature on public and private lands: a case study from Brisbane, Australia. Landsc Urban Plan 130(1). https://doi.org/10.1016/j.landurbplan.2014.06.005

Smiley KT, Sharma T, Steinberg A, Hodges-Copple S, Jacobson E, Matveeva L (2016) More inclusive parks planning: park quality and preferences for park access and amenities. Environ Justice 9(1). https://doi.org/10.1089/env.2015.0030

Swanwick C, Dunnett N, Woolley H (1978 Nature, role and value of green space in towns and cities: an overview. Environment 29(2). https://about.jstor.org/terms

Toftager M, Ekholm O, Schipperijn J, Stigsdotter U, Bentsen P, Grønbæk M, Randrup TB, Kamper-Jørgensen F (2011) Distance to green space and physical activity: a Danish national representative survey. J Phys Act Health 8(6). https://doi.org/10.1123/jpah.8.6.741

United Nations (2015) Transforming our world: the 2030 agenda for sustainable development

van den Berg M, Wendel-Vos W, van Poppel M, Kemper H, van Mechelen W, Maas J (2015) Health benefits of green spaces in the living environment: a systematic review of epidemiological studies. Urban Forest Urban Green 14(4). https://doi.org/10.1016/j.ufug.2015.07.008

van Dillen SME, de Vries S, Groenewegen PP, Spreeuwenberg P (2012) Greenspace in urban neighbourhoods and residents' health: adding quality to quantity. J Epidemiol Commun Health 66(6). https://doi.org/10.1136/jech.2009.104695

Venter ZS, Shackleton CM, van Staden F, Selomane O, Masterson VA (2020) Green apartheid: urban green infrastructure remains unequally distributed across income and race geographies in South Africa. Landsc Urban Plan 203. https://doi.org/10.1016/j.landurbplan.2020.103889

Wen C, Albert C, von Haaren C (2020) Equality in access to urban green spaces: a case study in Hannover, Germany, with a focus on the elderly population. Urban Forest Urban Green 55. https://doi.org/10.1016/j.ufug.2020.126820

Wickenberg B, McCormick K, Olsson JA (2021) Advancing the implementation of nature-based solutions in cities: a review of frameworks. Environ Sci Policy 125:44–53. https://doi.org/10.1016/j.envsci.2021.08.016

Winant H (2006) Race and racism: towards a global future. Ethn Racial Stud 29(5):986–1003. https://doi.org/10.1080/01419870600814031

World Health Organization (2016) Urban Green spaces and health: intervention impacts and effectiveness. www.euro.who.int

Wu L, Rowe PG (2022) Green space progress or paradox: identifying green space associated gentrification in Beijing. Landsc Urban Plan 219:104321. https://doi.org/10.1016/j.landurbplan.2021.104321

Xu M, Xin J, Su S, Weng M, Cai Z (2017) Social inequalities of park accessibility in Shenzhen, China: the role of park quality, transport modes, and hierarchical socioeconomic characteristics. J Transp Geogr 62. https://doi.org/10.1016/j.jtrangeo.2017.05.010

Yazdani S, Dola K (2013) Sustainable city priorities in Global North versus Global South. J Sustain Dev 6(7). https://doi.org/10.5539/jsd.v6n7p38

Young IM (1990) Justice and the politics of difference. Princeton University Press

Zhang R, Sun F, Shen Y, Peng S, Che Y (2021) Accessibility of urban park benefits with different spatial coverage: spatial and social inequity. Appl Geogr 135. https://doi.org/10.1016/j.apgeog.2021.102555

Investigating the Sustainability of Resource Flow and Productivity Transition in Ghana: Integrating MFA and DEA Based Malmquist Productivity Index Approach

Richard Agyemang, Liang Dong, Hatem Mahmoud, and Ibrahim Yahaya Wuni

Abstract In many developing countries, the efficiency of resource has become a prime blueprint target for suppliers of resource. This report examines resource productivity and efficiency in Ghana using the EW-MFA amalgamating with DEA Malmquist productivity index approach (DEA) (MDEA). In-depth analysis was done on the resource flow from 1978 to 2017 as well as the associated utilization efficiency taking unforeseen environmental effects into account. Over the past four decades in Ghana, DMC, GDP per capita, and material intensity (DMC/capita) indicated a relationship that is very strong from the results. The productivity of resource (USD/DMC) rise from 82.4 USD/ton to 125 USD/ton (2017) at a significant rate that is lower in Nigerian and Egypt. The increasing wealth (94%) and population (21%) based on the IPAT decomposition led to the rapid explosion of the DMC, while improvement

R. Agyemang (✉)
Institute for Infrastructure and Resource Management, Faculty of Economics and Management Science, University of Leipzig, Saxony, Germany
e-mail: richardagyemang245@gmail.com

L. Dong
Department of Public Policy, School of Energy and Environment, City University of Hong Kong, Tat Chee Avenue, Kowloon, Hong Kong, China
e-mail: liadong@cityu.edu.hk

H. Mahmoud
Departments of Architecture, Faculty of Engineering, Aswan University, Aswan, Egypt

Department of Environmental Engineering, Egypt Japan University of Science and Technology, Alexandria, Egypt

H. Mahmoud
e-mail: hatem.mahmoud@ejust.edu.eg

I. Y. Wuni
Departments of Construction, Engineering and Management, King Fahd University of Petroleum and Minerals, Dhahran 31261, Saudi Arabia
e-mail: Wuni-ibrahim.wuni@kfupm.edu.sa

A. Sharifi et al. (eds.), *Bridging Peace and Sustainability Amidst Global Transformations*, World Sustainability Series,
https://doi.org/10.1007/978-981-99-7572-3_10

needed to be done by the technology element (DMC/GDP).The total factor productivity comparison among Ghana, Egypt, South Africa, and Nigeria revealed that, on the one hand, Ghana's economy has seen a significant upward trend with EFFCH 1.061 and TECHCH 1.046 during the past four decades. However, Ghana continues to be a low-productivity, material-intensive country. By optimizing resource usage, Ghana can alleviate strains on their environment, contributing to greater harmony within ecosystems. Moreover, as resource efficiency leads to economic growth and stability, it fosters the foundation for peaceful coexistence by addressing socio-economic disparities. The pursuit of sustainable practices, driven by technological innovation and resource optimization, thus paves the way for a more peaceful and prosperous future for nations and the global community as a whole.

Keywords Resource metabolism · MFA · Dematerialization · IPAT · DEAM

1 Introduction

The UN Sustainable Development Goals must be achieved by 2030, which calls for resource efficiency and sustainable resource management. In the near future, higher raw material prices and a resource deficit will be caused by increased resource consumption and materialization for economic growth (Schandl and Turner 2009); and many nations strengthen material efficiency measures in their economic systems (Giljum et al. 2014a). Resource efficiency can be simply explain as "lower resource use and associated environmental impacts with long-term economic advantages" (Schandl and West 2012; Shah and Park 2020) in order to arrive at a development that is completely balanced. As a result of resource overuse, the ecosystem's limited ability to absorb waste as well as its ability to regenerate itself has both been exceeded.

A particularly severe effect of over utilization of resource environmental variation (Chika et al. 2012; Wiedenhofer et al. 2019), such as increased in atmospheric temperature (global warming) and deterioration of resources that are provided by nature. The efficiency and productivity of resource are vital in particular to nations that serve as the major distributors of resource of industrial products if SDGs is to be achieved at the national level. Deeper investigation of material metabolism is a crucial choice that will enable sustainable resource management. An important analytical tool in industrial ecology known as material flow analysis (MFA) balances the system's resource flows and offers crucial insights into the system's long-term resource management. MFA rules have been issued by Eurostat (Eurostat 2001, 2017, 2013). Studies on the scope of national, regional, municipal, urban and industrial have been conducted using MFA. MFA (EW-MFA) on national level is the most famous and modern approach for national accounting on the social and economic metabolism (Dong et al. 2017). Resource efficiency and use was analyzed using EW-MFA as well as improvement based on support planning which has a significant impact to the nation's sustainable resource development and management. EW-MFA studies have been carried out in countries such as those that are EU members (Balat 2004), Hungary (Hammer and Hubacek 2003), Australia (Schandl and West 2012), Pakistan (Shah and Park 2020), and nations that are grouped by region (Calvo et al. 2016; Giljum et al. 2014b; De Castro Camioto et al. 2016). The efficiency of an economic- ecosystem has currently been used by MFA based on Data Envelopment analysis (DEA), which quantifies a system's efficiency by the ratio of weighted outputs to inputs (Shah et al. 2020). Socioeconomic factors can help in the comprehension of resource management based on the resource flow and production efficiency as a crucial role for long-term development at the national or regional level (Shah and Park 2020; Dong et al. 2017).Ghana's outstanding successes in economic transformation have earned it the reputation of being among the nations that are developing the fastest. The nation did remarkably well in 2017, with a GDP gain of 6.4%. At the time, it was anticipated that the world GDP would expand by about 3% (Update and Opment 2012). There is an evidence of increasing challenges with great resource needed and emissions of carbon due to the increased in DMC from 1990 to 2016 (from 16.06 Mt to 61.41 Mt. there was also an increased in emission intensity from 4.75 to 5.12 Mt CO_2 per ktoe of fossil fuels (Huong et al. 2021). Furthermore, Ghana's Ministry of Natural Resources and Environment projects a 20% rise in carbon intensity per GDP and a fivefold increase in greenhouse gas emissions. These statistics back Ghana's pro-growth policy, which places a higher value on economic assets like resources and a clean environment than on environmental assets like resources and a clean environment (Bass et al. 2009), leading to rapid resource depletion and environmental deterioration as well as poor resource and environmental management. Much study has not been conducted with regards to productivity and the movement of material in the country Ghana. In order to deeply derive mush insight on Ghana's transitional features of economic explosion, utilization of resource, efficiency, and pollution emissions, there is the need for research studies on the flows of material and the efficiency in order to finalize policies of sustainable development merging environmental and development objectives. In light of these circumstances, this report investigate the sustainability of resource flow and productivity transition in Ghana from 1978 to 2017 using an analysis of economic-wide material flows and a DEA-based Malmquist Productivity Index approach (denotes DEAM). Due to geographical and trade proximity, Ghana is compared with other countries to examine how material and carbon efficiencies have changed over.

Our knowledge and a review of the literature indicate that no published studies on this subject using comparable methodologies exist. These research inquiries are addressed in this paper: (1) What factors influence material consumption in the economy, (2) how has material consumption changed in Ghana, a country with rapid economic growth, and (3) how have material and carbon efficiencies changed through time at the national level in comparison to Egypt, South Africa, and Nigeria. We intend to address research issues and create significant policy suggestions for Ghana's material consumption, productivity, and long-term development from a historical and macro-policy viewpoint. By delving into these aspects, this research endeavor not only addresses critical research questions but also contributes to policy recommendations for Ghana's sustainable development journey. In this way, the study directly aligns with the broader goal of achieving peace and sustainability. It recognizes that a harmonious future requires equitable resource management, responsible economic growth, and a keen understanding of the intricate interplay between resource consumption, efficiency, and environmental well-being.

Following the introduction, Sect. 2 uses Ghana as a case study to analyze the overall situation. Section 3 then presents the methodology and data. Section 4 then presents the analytical results and leads a discussion. Finally, Sect. 5 comes to a conclusion.

2 Overview of Ghana

Ghana is a country on the West of Africa with a populace of approximately 31 million and represents 0.4% of the global population. In 2020, the number of females was 15.32. Million whereas that of males was15.75 Million. The country has 6 major ethnic groups with the largest tribe being the Ashanti, which has it capital in Kumasi. Ghana's neighboring countries includes Burkina Faso, Togo, La Cote d'Ivoire and the Ocean Atlantic. The major economic activity is dedicated to the Agricultural sector which employs majority of the populace and also a major contributor to the country's GDP from 2010 to 2019. The country is also rich in minerals like Gold, Diamond, Bauxite, and Manganese. Natural resources serve as the foundation for the majority of the nation's export goods. Starting in 2011, these include cocoa, gold, and most recently, crude oil. Over the past three years, the production of natural gas and crude oil exports in particular has contributed to accelerating economic growth. In Ghana, carbon dioxide continued to be the primary GHG, making up around 66% of all net emissions in 2016.

2.1 Economic Development Status of Ghana

Ghana has the second-largest economy in West Africa as a result of its robust exports of oil, gold, and cocoa. The country Ghana is considered as oldest democracies in Africa which has a long history of democratic elections and variations in political power among parties.

The country began the production of crude oil in 2011 at the Jubilee, and this has led to a tremendous increased in the country real GDP by 15% in 2011 and 7.9% in 2012. Other sectors such as gold and cocoa also contributed to this increased. The county can confidently conclude that by expanding service of the economy this trend increasing may continue and will be very instrumental pro the economic expansion of the nation. The lockdown in the country as spearheaded by the government of Ghana in March 2020 due to the COVID-19 outbreak led to a reduction in the exports of commodities which slowed Ghana's rapid development. The economy of Ghana increased at an annual rate of 7% rate (2017–2019) before bearishly declining in 2020(second and third quarters.

The economic downswing had a tremendous effect on households. The poverty rate is expected to rise from a percentage of 25 to 25.5 in 2019 and in 2020. Also, by 2020, the country deficit budget will have increased to 15.2% of GDP. The country public debt has currently exposed the country into the risk of challenges in finance as the debt has risen to 81.1% of GDP in 2020. Irrespective of a substantial downfall attributed to the mining sector and the second wave of the pandemic, in the first and second quarters of 2021, there was massive in the acceleration of growth. The government authorities have made it a point to reduce expenses to make up for the reduction of revenue for the initial half of 2021 based on preliminary financial data. 5.1% of GDP constituted the fiscal deficit of the country. The inflation was at 7.8% in June 2021 which continued to be low as the COVID-19 as the prices of food subsided. Due to this, the Bank of Ghana reduced it rate of policy in May by 100 basis points to 13.5% to help in the redemption of the economy. In August, there was 9.7% increased in inflation due to increasing food prices and non-food inflation (Fig. 1).

Fig. 1 Map showing Ghana and its neighboring countries

2.2 Resource Consumption Patterns

Figure 2 shows how resource consumption in Ghana increased in lockstep with economic growth as indicated by domestic material consumption (DMC) between 1978 and 2017 (Schandl and Turner 2009).

In 2017 from the graph, there was an increased in the final DMC to 200.47 Mt (17.6 times) from 35.80 Mt. There was an improvement in the standard of living during the preceding 40 years (files 2017). There was also a variation in the structure indicating a colossal expansion in the economy as well as industrial explosion. In 1978, the DMC

Detail information (Figure 3)		1978	2017
Biomass (Mt)	Crop Residues	16.81	16.7.3
	Crops	19.72	22.0.9
	Grazed biomass and fodder crops	5.09	1.42
	Wild catch and harvest	0.80	0.14
	Wood	17.7	14.3
Non-metallic minerals (Mt)	Non-metallic minerals - construction dominant	18.75	23.66
	Non-metallic minerals - industrial or agricultural dominant	6.61	0.13
Fossil fuels (Mt)	Oil shale and tar sands	0.00	0.00
	Petroleum	0.00	2.88
	Coal	0.00	00.00
Metal ores (Mt)	Ferrous ores	0.00	1.42
	Non-ferrous ores	10.93	14.87
			11

Fig. 2 During the period 1978–2017, resource consumption (DMC) by material type was calculated (proportional DMC values are in percentages)

of the industry, construction, nonmetallic mineral consumption gradually increased from 37.9 to 70.91%. The non metallic mineral category recorded an increase from 19.4% in 1978 to 27.64% in 2017 as a result of the industrial resolution and the rising economic affluence. Due to the reduction in the consumption of other resources, the percentage rise from −9.18 to 6.92%. On the contrary, the amount of biomass used in total DMC increased from −0.12 Mt in 1978 to 0.98 Mt. (2017). Ghana's transitioning policy of rural economy into a modernized and an industrialized economy is in adherence to the recent dimensions of consumption of material (Barker and Üngör 2019). There was an increased in the consumption of fossil fuel in 1978 from 0.92 to 8.89 million tons in 2017(by about 20 times). Nevertheless, fossil fuels percentage in DMC declined from 71.14 to 22.75% with respect to those years. The quantum of metal ores exploded from decreased 0.85 to −0.59 in the DMC. The rising economic affluence, improvement in the standard living of people as well as the industrial revolution in Ghana, the consumption of construction materials and fossil fuel is expected to indefinitely arise. However, Ghana must also encourage the importation of resources from other nation if the extraction of domestic resources fails to comply with the DMC demand (Shah and Park 2020). Owning to this, there is the need for an efficient and effective management of resource policies and principles in relation to the flow of resource and productivity efficiency conversion.

3 Methods and Data

The approach and the required data employed in this study are both evident in Fig. 3.

Resource groped into four was employed in classifying materials into 13 distinct categories (metal ores, fossil fuels, non-metallic minerals, and biomass). The first step is to conduct a material flow analysis for Ghana's MFA from 1978 to 2017 based on current long-times series data traversing 4 decades. The EKC hypothesis is then applied in the second stage, using decomposition analysis to determine the factors that influence resource utilization. To measure economic success, the Malmquist productivity index technique based on DEA is utilized (DEAM). Ghana is likened to the most developed African economies since material metabolism is impacted by economic growth stages (Nigeria, Egypt, and South Africa).

3.1 Material Flow Analysis

The study complies largely with the EW-MFA composition and approach criterion in EUROSTAT (Eurostat 2001, 2017, 2013) and Table 1 give an account of the MFA indicators. MI, EI, and RP which depicts the indicators of intensity and efficiency are conveyed in mixed units (DMC/capita and/or DMC/GDP, CO_2/capita and/or CO_2/GDP, and GDP/DMC) whereas physical units (Mt) are used in expressing indicators like DE, DMC, and PTB. These MFA indicators are used to look into trends in

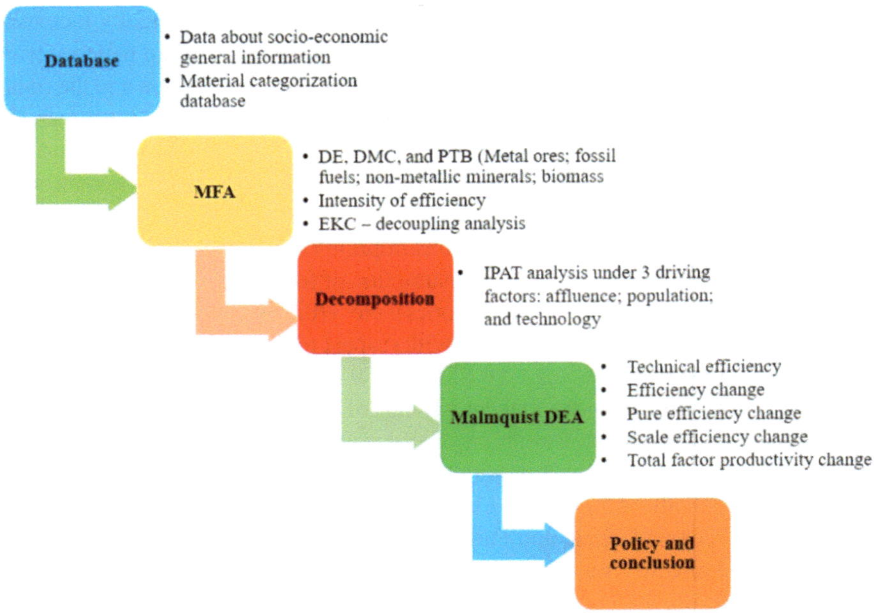

Fig. 3 MFA and DEAM approach and data for the study

resource productivity and dematerialization/decoupling (Pivnenko et al. 2016; Kósi and Torma 2005). 2010 constant prices (or otherwise stated) in US dollars (USD) in relation to GDP is employed as performance economic indicator was used in the analysis.

Table 1 Material flow and efficiency indicators applied in this study

Indicator	Abbreviation	Description
Domestic extraction	DE	Resources taken from the natural world and exploited as a source of income (except water and air)
Domestic material Consumption	DMC	The total amount of domestically produced raw materials consumed in an economy, including imports and minus exports
Physical trade balance	PTB	Material resource trade surplus or deficit, calculated as physical imports minus physical exports
Intensity	MI EI	The total amount of material directly consumed in an economy, as determined by DMC per capita and DMC per GDP, is represented by total CO_2 emissions/capita or total CO_2 emissions/GDP
Resource productivity	RP	Resource efficiency is measured using the GDP/DMC ratio, which is based on the ratio of GDP to DMC

3.2 IPAT Decomposition Analysis

IPAT (Nakicenovic and Swart 2000; York et al. 2003) is a straightforward equation that asserts that population (P), affluence (A), and technology (T) are the multiplicative products of three socioeconomic elements that have an impact on the environment (T). It is used along with the MFA data to delve deeper into the factors that influence resource utilization. Under the IPAT assumption, DMC was used to represent the environmental impact (I) caused by the three principal causes listed in Eq. (1) $I = P \times A \times T$:

$$Env.Impact = Population \times \frac{GDP}{Capita} \times \frac{Env.Impact}{Unit\,of\,GDP} \tag{1}$$

When P denotes the total number of people, A denotes the level of prosperity (GDP per person), and T denotes technology (DMC per GDP). To quantify the contributions of the three factors, the Log-Mean Divisia Index (LMDI) technique (Ang and Liu 2001) was employed to analyze the change in DMC from "time = 0" (the start year) to "time = t" (the end year) (the end year). Using the LMDI approach, the difference in DMC from "time = t0" to "time = t" can be expressed as the following equation: $\Delta I(env.impact) = DMC(tons) = DMC_{t1}(tons) - DMC_{to}(tons)$

$$= \Delta P + \Delta A + \Delta T \tag{2}$$

$$\Delta P = \sum \frac{DMC_{t1} - DMC_{t0}}{InDMC_{ti} - InDMC_{to}} \times In\frac{P_{t1}}{P_{t0}} \tag{3}$$

$$\Delta A = \sum \frac{DMC_{t1} - DMC_{t0}}{InDMC_{ti} - InDMC_{to}} \times In\frac{A_{t1}}{PA_{t0}} \tag{4}$$

$$\Delta T = \sum \frac{DMC_{t1} - DMC_{t0}}{InDMC_{ti} - InDMC_{to}} \times In\frac{T}{T_{t0}} \tag{5}$$

where P, A, and T denote the population, affluence, and material intensity contributions, respectively.

3.3 Data Envelopment Analysis-Based Malmquist (DEAM) Approach

Using the MFA indicator and Data Envelopment Analysis, we first attempt to evaluate the economic system's productivity in this paper (DEA). A non-parametric mathematical technique called DEA is used to assess the effectiveness of systems at a specific moment in time. The productivity efficiency index for organizational units can be evaluated over time using the Malmquist Productivity Index technique

(MDEA), which is based on data envelopment analysis (Zhang et al. 2021). Decision-making units (DMUs) are the organizational entities that convert inputs into outputs (Charnes and Cooper 1984; Wang et al. 2019; Liu et al. 2015). Numerous performance (efficiency) issues have been resolved with DEA (Zhang et al. 2021; Lozowicka 2020; Li et al. 2020; Chang et al. 2019; Afsharian and Ahn 2015; Long et al. 2020). However, there is a glaring lack of national-level research on changes in resource productivity efficiency in the context of economic growth and environmental impact, which could offer helpful insights into sustainable resource management policies. This is especially true in emerging and developing economies. In light of this, this study was the first to assess Ghana's productivity efficiency changes in comparison to the continent's main economies using MFA indicators and the DEA-based Malmquist productivity Index approach.

Equations (6) and (7), which are calculated using the DEA distance function to determine the efficiency change of the DMU between periods t and t + 1, are known as the Malmquist productivity index (MPI) (Roh et al. 2011).

$$At\, time\, t: \quad MPIt^{t}_{I,c} - \frac{Hc^{t}(x^{t+1,y^{t+1}})}{HI_{c}(x^{t}, y^{t})} \tag{6}$$

$$At\, time\, t+1: \quad MPIt^{t+1}_{I,c} = \frac{Hc^{t+1}(x^{t+1,y^{t+1}})}{H^{t+1}(x^{t}, y^{t})} \tag{7}$$

While Ht (xt, yt) and Ht (xt + 1, yt + 1) denote the distance function of input and output of DMUs by period t benchmark technology, respectively, the Malmquist productivity index (MPIt (I,c)), which is the geometric mean of two MPI, measures Total Factor Productivity (TFP). Where I denotes the orientation (input or output orientation); C denotes technology under constant return to scale; MPI (t $MPIt^{t}_{I,c}$ $(\boldsymbol{x}^{t}, \boldsymbol{y}^{t}, \boldsymbol{x}^{t+1}, \boldsymbol{y}^{t+1}) =$

$$\left(MPIt^{t}_{I,c}MPIt^{t+1}_{I,c}\right)\frac{1}{2} = \left(\frac{Hc^{t(x+1,y+1)Hc^{t+1}x^{t+1,y^{t+1}}}}{HI_{c}(x^{t}, y^{t})Hc^{t+1}\left(x^{t+1,y^{t+1}}\right)}\right)\frac{1}{2} \tag{8}$$

If MPIt (I,c)t(xt, yt, xt + 1, yt + 1) > 1, TFP grows from t to t + 1, and if MPIt (I,c)t(xt, yt, xt + 1, yt + 1) > 1, TFP declines. The Eq. (8) is decomposed further to give Eq. (9)

$$\begin{aligned} MPIt^{t}_{I,c}(\boldsymbol{x}^{t}, \boldsymbol{y}^{t}\boldsymbol{x}^{t+1}, \boldsymbol{y}^{t+1}) &= \left(\frac{Hc^{t(x+1,y+1)Hc^{t+1}x^{t+1,y^{t+1}}}}{HI_{c}(x^{t}, y^{t})HI_{c}(x^{t}, y^{t})}\right) \\ \frac{1}{2}\left(\frac{Hc^{t(x+1,y+1)}}{HI_{c}(x^{t}, y^{t})HI_{c}}\right)&\left(\frac{Hc^{t(x+1,y+1)Hc^{t+1}x^{t+1,y^{t+1}}}}{Hc^{t+1}\left(x^{t+1,y^{t+1}}\right)(Hc^{t+1}(x^{t,y^{t}})}\right)\frac{1}{2} \end{aligned} \tag{9}$$

The following categories under which technology change (TECHCH) and efficiency change (EFFCH) can be classified

$$\text{CRS assumption}.TECHCH = \left(\frac{Hc^{t(x+1,y+1)}Hc^{t+1}x^{t+1,y^{t+1}}}{Hc^{t+1}\left(x^{t+1,y^{t+1}}\right)\left(Hc^{t+1}\left(x^{t,y^{t}}\right)\right.}\right)\frac{1}{2} \tag{10}$$

$$EFFCH = \frac{Hc^{t+1}\left(x^{t+1,y^{t+1}}\right)}{Hc^{t+1}\left(x^{t,y^{t}}\right)} \tag{11}$$

Therefore,

$$\text{TFPCH} = \text{TECHCH} \times \text{EFFCH} \tag{12}$$

Equations 1 and 2 are further divided into efficiency changes (EFFCH) in accordance with the variable returns scale (VRS) assumption (13). Equation (13) has two parts: scale efficiency change (SECH, equation) and pure efficiency change (PECH, Eq. 14).

$$\frac{Hc^{t+1}\left(x^{t+1,y^{t+1}}\right)}{Hc^{t+1}\left(x^{t,y^{t}}\right)} = \frac{Hc^{t+1}\left(x^{t+1,y^{t+1}}\right)}{Hc^{t+1}\left(x^{t,y^{t}}\right)}\left(\frac{\frac{Hc^{t+1}\left(x^{t,y^{t}}\right)}{Hc^{t+1}\left(x^{t,y^{t}}\right)}}{Hc^{t+1}\left(x^{t+1,y^{t+1}}\right)Hc^{t+1}\left(x^{t+1,y^{t+1}}\right)}\right) \tag{13}$$

$$PECH = \frac{Hc^{t+1}\left(x^{t+1,y^{t+1}}\right)}{Hc^{1}\left(x^{t,y^{t}}\right)} \tag{14}$$

$$SECH = \frac{Hc^{t}\left(x^{t,y^{t}}\right)/Hc^{t+1}\left(x^{t,y^{t}}\right)}{Hc^{t+1}\left(x^{t+1,y^{t+1}}\right)/Hc^{t+1}\left(x^{t+1,y^{t+1}}\right)} \tag{15}$$

Therefore, TFP change can be written as follows:

$$\begin{aligned}\text{TFPCH} &= MPIt_{I,c}^{t}(\mathbf{x}^{t}, \mathbf{y}^{t}, \mathbf{x}^{t+1}, \mathbf{y}^{t+1}) \\ &= \text{TECHCH} \times \text{EFFCH} \\ &= \text{TECHCH} \times \text{PECH} \times \text{SECH}\end{aligned}$$

The efficiency frontier and efficiency score in traditional DEA are determined by inputs (such as energy, material, and capital) and outputs (such as GDP, value-added, and waste emissions) (distance function). We use material intensity (DMC per Mil \$) and carbon intensity (CO_2 per Mil \$) (unwanted production) as inputs to be minimized and economic growth (GDP Mil \$) as an outcome to calculate productivity efficiency scores (Shah and Park 2020).

3.4 Data Collection

General information on socioeconomic and environmental issues was available from the World Bank (The World Bank 2017; World Bank 2019) and the International Energy Agency's database (World Bank 2019). The International Resource Panel's database, which was based on the EUROSTAT architecture (Eurostat 2001, 2017, 2013), provided the consumption information for material classification. For the policy study, publically available policy documents were accessed and evaluated.

4 Results and Discussion

This section gives a detailed discussion of the results of the studies based on the method incorporated in the analysis.

4.1 Transition of Basic Material Indicators

Metal ores and industrial minerals, fossil fuels, and building supplies, along with biomass, make up the three main components. Domestic basic material extraction per person is shown in Fig. 4a, while domestic material consumption and physical trade balance statistics are shown in Fig. 4b, c for the years 1978–2017.

Figure 4 shows how each material type's share and the DE trend have drastically altered over time (a). Total DE in Ghana has climbed from 34.51 Mt to 193.57 Mt over the past 40 years at a compound annual growth rate (CAGR) of 4.52%. An increase in the share of construction minerals in total DE from 37.11% in 1978 to 67.90% in 2017 indicates that extraction of construction-based materials is predominate. In 1978, biomass accounted for 21.87 Mt of total DE, or 57% of all DE. In 2017, biomass accounted for 83 Mt of total DE (193.57 Mt), or 111.3 Mt. Construction minerals were therefore the resource that was most heavily utilized in the Ghaana as of 2017, with the largest increase occurring between 1978 and 2017.

Over the study period, the overall share of fossil fuels in DE increased moderately, following a similar pattern to the overall share of biomass (0.92 Mt in 1978 to 4.37 Mt in 2017). This is mostly due to a considerable rise in the economy's share of construction materials. Construction materials have been employed to build industrial process facilities (for industrialization) and extensive urban infrastructure over the previous 40 years of economic growth (for urbanization). Meanwhile, the extraction of fossil fuels increased from 0.92 Mt in 1 to around 10 times in 2017 (Fig. 4), reflecting the importance of fossil fuels as energy resources, particularly for industrialization, urbanization, and the improvement of living standards (a).

Ghana's domestic material consumption (DMC) per person is shown in Fig. 4b over the period of 1978–2017. With a CAGR of 4.52%, total DMC expanded from

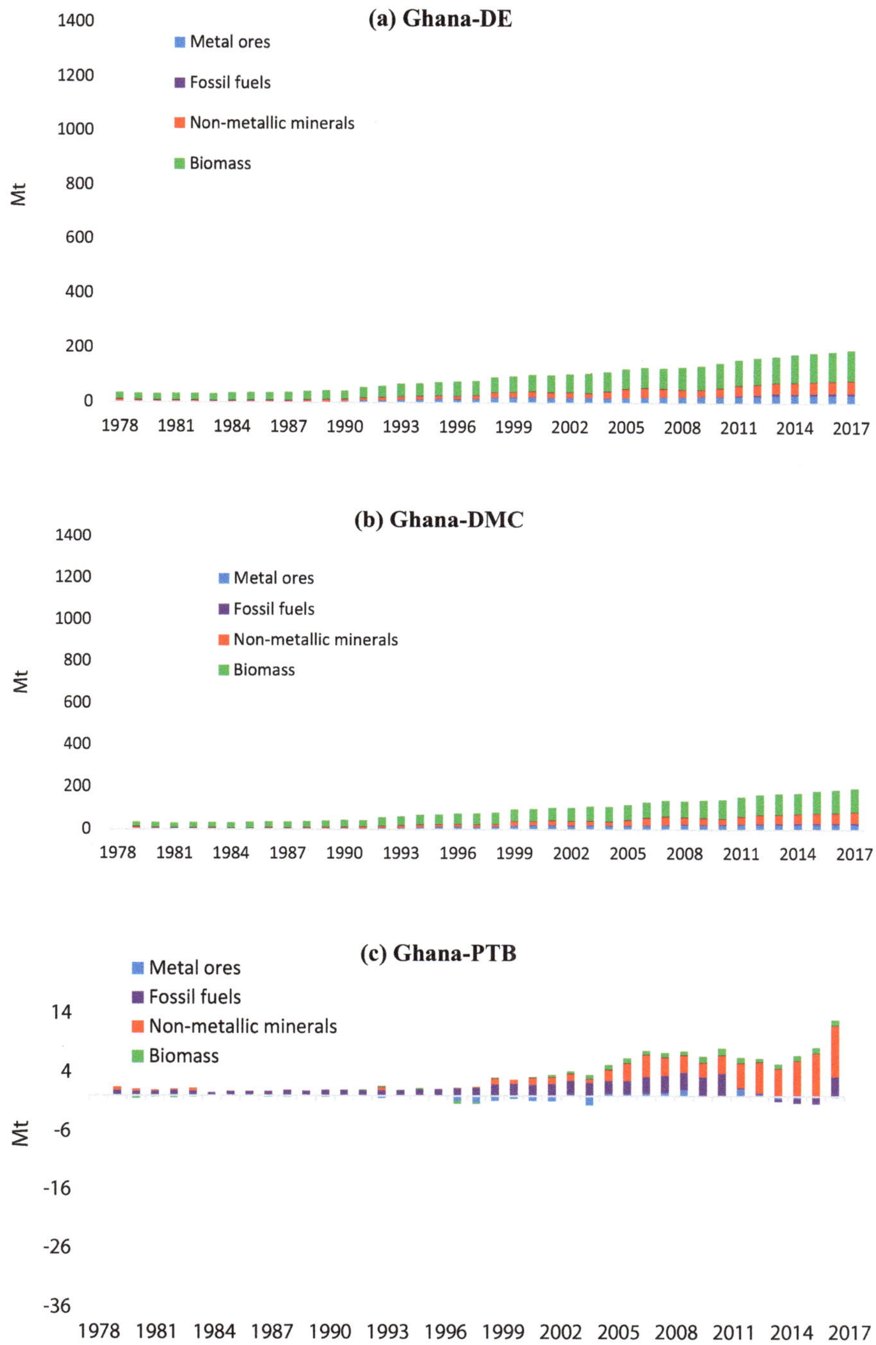

Fig. 4 **a** Domestic extraction (DE), **b** domestic material consumption (DMC), and **c** physical trade balance (PTB) in Ghana during 1978–2017

38.50 Mt in 1978 to 200.47 Mt in 2017, primarily due to rapid advancements in economic growth, industrialization, and urbanization. In relation to locally mined or imported materials, resource utilization is shown by the DMC to DE ratio. The DMC/DE ratio was approximately 1.0 in 1978, and it stayed that way until 2017, showing Ghana's comparatively low reliance on imported resources. This ratio of metal ores to fossil fuels was larger than 1, 2.5, and 1.1 as of 2017, indicating that imports of those commodities were considerable because to DE shortage. Ghana's growing reliance on imported resources is developing as a significant barrier to economic advancement given the complexity of future global resource supply chains, resource price volatility, and expected competition among importing-trading countries.

A positive PTB indicates net resource import and vice versa, and Fig. 3c shows Ghana's PTB trajectory from 1978 to 2017, which offers critical insights into net resource flows to and from an economy (Shah and Park 2020). Three fascinating phases are shown in this figure: 1978–1988, 1989–1999, and 2000–2017. In order to overcome the repercussions of violence, social development and reconstruction were the main national goals from 1978 to 1988. In order to address the nation's basic needs during this time, the necessary raw materials and resources were prioritized for import. However, Ghana's net resource imports increased from 1.0 Mt to 1.5 Mt, remaining relatively low and largely consisting of imports of fossil fuels (1.5 Mt). Ghana's natural resource exports surpassed its imports from 1989 to 1999, and exports of cheap raw materials are essential for the country's economic development.

Between 2000 and 2017, the import and export dynamics of raw materials were clearly discernible. Due to rising domestic demand, a shortage of fossil fuels was created in Ghana in 2017 as a result of a sharp increase in imports (−5.9%). Metal ore imports consistently rose over nine times over this time, from 20.19Mt in 2000 to 31.46 Mt in 2017, to meet the demands of industrial activities. Between 2000 and 2009, biomass was imported; beginning in 2011, it was exported. From 2000 to 2017, more metallic ores were imported than exported to finance the construction of urban and industrial infrastructure. Non-metallic minerals have been rapidly explored and have increased almost threefold between 2010 and 2017 to become one of the major resources exported from this nation. According to projections, these resource PTB dynamics will persist in response to rising living standards, expanded urban infrastructure, and quickening industrial growth, demonstrating an increasing sensitivity to disruptions in the supply of resources globally and changes in their prices (Shah et al. 2020).

4.2 Material Flow and Intensity Trends

The fundamental material flow indicators, intensity, and efficiency developments in Ghana between 1978 and 2017 are examined, as well as the top three economies in Nigeria, Egypt, and South Africa (Fig. 5). The figure shows GDP per capita, resource productivity in DMC tons per capita, resource productivity in GDP per DMC, and environmental effect in CO_2 per capita as measures of economic growth.

Nigeria's GDP per capita has increased 23.9 times in a row in terms of economic growth. However, over the course of four decades, material intensity (DMC per GDP) has decreased in Nigeria, Egypt, and South Africa (Fig. 5a), whereas in Ghana, its trajectory has fluctuated, rising from 0.0009 tons per US dollar in 1978 to 0.0118 tons per dollar in 1994 before falling to 0.0036 tons per US dollar in 2017. This reveals that Ghana continues to be the most materially intensive country overall, despite significant advances, and has the lowest productivity. A crucial factor in determining the economic contribution per unit of material used is the economic production per unit of total DMC, which is represented as resource and material productivity. Figure 5c shows that at that time, material productivity climbed 1.5 times in Ghana, going from 111.62 to 271.85 USD/ton, whereas it increased 164, 28 and 335 times in Nigeria, South Africa, and Egypt, respectively. In 2017, Ghana was behind Nigeria, South Africa, and Egypt in terms of material productivity (564, 579, and 417t, respectively).

Figure 5d illustrates how Ghana's ton CO_2 per capita grew 11 times between 1978 and 2017 based on carbon intensity, compared to increases of 9 times, 12 times, and 13 times in Nigeria, South Africa, and Egypt over the same time period. Egypt had the greatest carbon intensity growth rate in 2017, despite Nigeria's (0.8485) and South Africa's (1.0249) rising rates being lower than Ghana's (2.5984). They must therefore enhance climate mitigation strategies including the circular economy, green growth, and the promotion and execution of sustainable development. Ghana's carbon intensity is relatively low, but the nation is steadily moving in the direction of low-carbon economic growth thanks to significant advancements in energy efficiency and emission intensity as well as changes to the country's overall energy mix (Huong et al. 2021). In summary, trends in intensity and efficiency show that patterns of resource consumption and economic development vary between nations, depending on factors like the state of the nation's resources, stage of economic development, industrial structure, and position in global supply chains [58]. To obtain crucial policy insights on sustainable resource management for the target country, nation-specific MFA is necessary (Schandl and West 2012; Wang et al. 2012).

4.3 Coupling Trends and Drivers of Resource Consumption

The material intensity and economic variables acquired from MFA were utilized to evaluate decoupling trends using the Environmental Kuznets Curve approach. Figure 6 shows how Ghana and Nigeria have been in a downward spiral in terms of GDP per capita and resource consumption. These two countries fall into the initial portion of the inverse "U" curve and have shown a negative link between DMC per capita and economic growth during the past 40 years. In contrast, Nigeria showed transitional tendencies and reached the inversion point in 1987 with a per capita GDP of 32,000 USD. The decoupling of material intensity from economic growth is evident here. This finding suggests unequivocally that Ghana, Egypt, and South Africa must advocate for policies that support resource efficiency, green growth, and

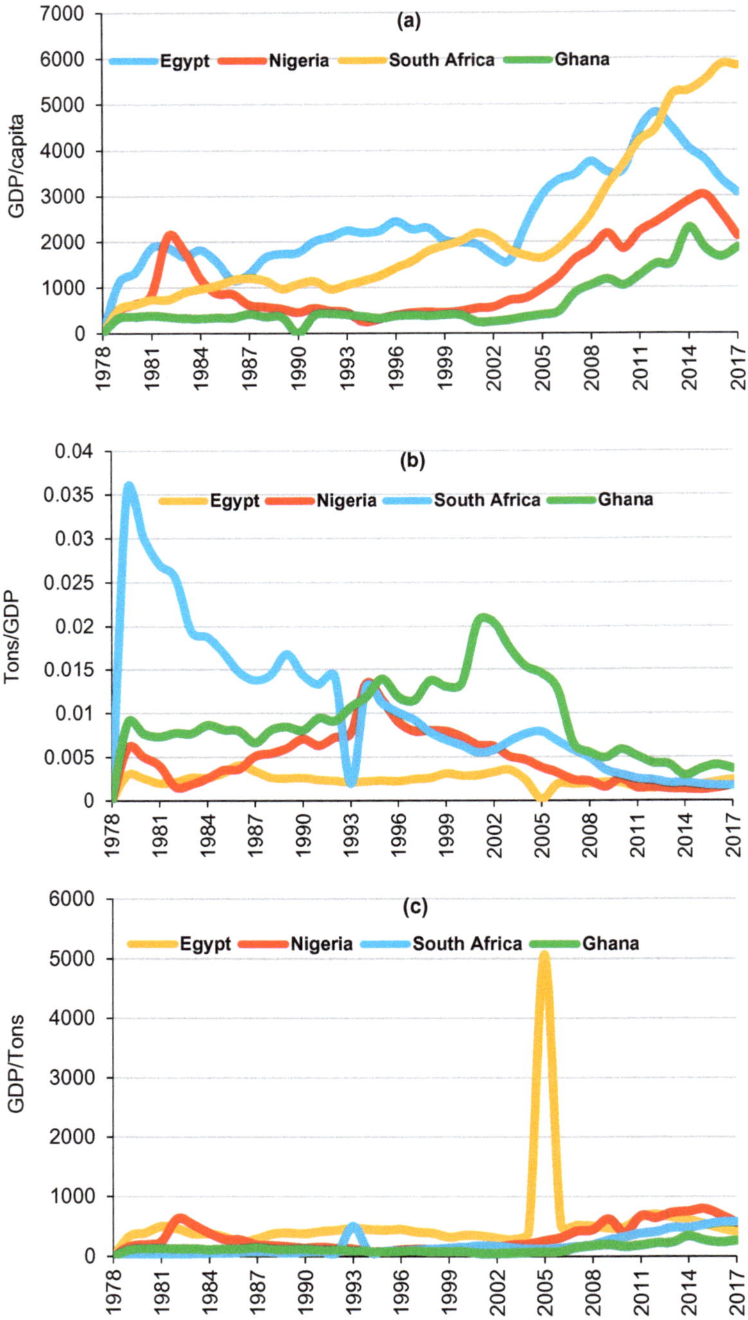

Fig. 5 Ghana, Nigeria, South Africa, and Egypt's economic development, resource use, and emissions from 1978 to 2017. GDP is expressed in constant 2010 dollar prices. GDP per person, resource productivity (GDP/DMC), resource intensity (DMC/GDP), and environmental effects are only a few examples (CO_2)

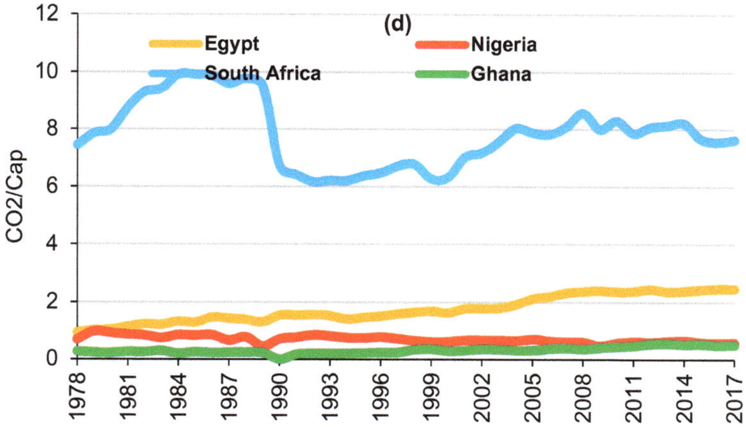

Fig. 5 (continued)

circular economies and take decisive action to raise material intensity and material productivity. Though Ghana is still far from the EKC inversion point, it is strongly advised that nation strengthen technical innovation (a negative driver) catered to local conditions in order to decouple resource consumption from economic growth given the significance of technical factors in material consumption.

The results for Ghana and comparative nations are shown in Table 2, which deconstructs DMC into its driving variables every 10 and 5 years in order to analyze resource consumption's motivators in greater detail (Two important African

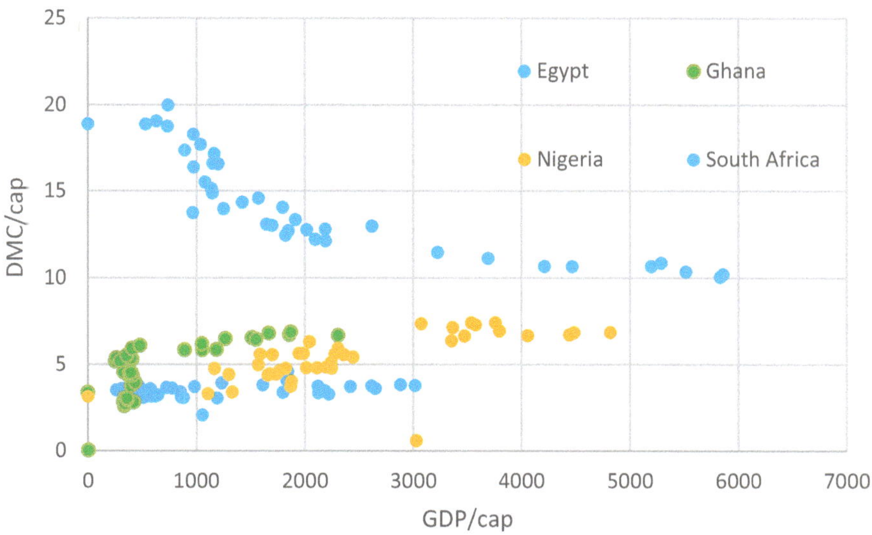

Fig. 6 EKC curve of Ghana comparison with Nigeria, South Africa, and Egypt during 1978–2017

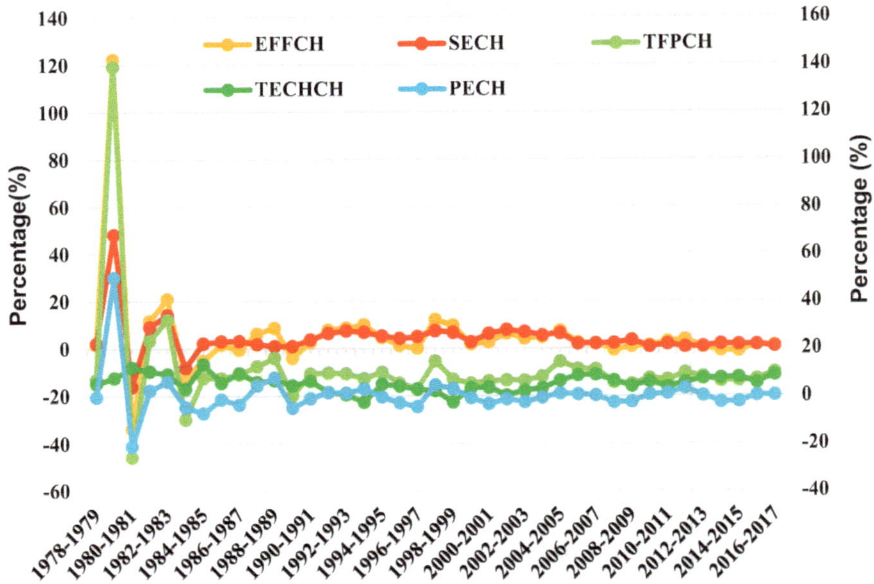

Fig. 7 EFFCH, PECH, and SECH from 1978 to 2017. Note that TFPCH and EFFCH values are shown on the second axis. Figure 8 illustrates the TFPCH from a national perspective together with its material consumption and eco-efficiency components. The average yearly TFPCH was 1.097, as shown in Table 3

economies). While population and economic expansion have a negative influence on material consumption over the research period, technological advancement has a positive impact on resource use and environmental problems. A growing population and increased industrial activity inevitably require basic essentials for survival and productivity. Because both contributions have a disproportionately large impact on material and environmental problems. Ghana, Nigeria, and Egypt had populations that indicated increased material consumption and emissions throughout the four decades under study at rates of 197%, 103%, and 44%, respectively (Table 2). In Ghana, Nigeria, and Egypt, respectively, it contributed 197, 315, and 149% to affluence between 1978 and 2017. Innovative approaches, on the other hand, are thought to be a major cause of Nigeria's declining resource and environmental conditions (−189% contribution of technological factor during 1978–2018). Egypt contributed −94% of this effort, which was followed by China with a −146% share of technical advancement for the entire period. As shown, increasing income (197%) and population growth (59%) are the main forces behind DMC, with only technological causes contributing to a reduction in material consumption in Ghana (−156%) between 1978 and 2017 (Table 2). In particular, the technology factor was the negative driver of DMC and significantly slowed the rise in material consumption during 1978–1988, the first phase of economic institutional transition (−427%), whereas the affluence factor and population driver both markedly increased resource consumption (330% and 197%, respectively) (Table 2). The DMC was 34.03 in 2013 and 28.39 in 2017 as

Table 2 Driver factors of material consumption in Ghana under intervals

Change in impact (DMC)			Change in impact (DMC)			Attribution to drivers using log transform		
	ΔI%	ΔI (Mt)	ΔP (%)	ΔA (%)	ΔT (%)	P (%)	A (%)	T (%)
1978–1988								
Ghana	15	5.37	32	59	−45	197	330	−427
Nigeria	29	65.37	31	192	−66	103	415	−418
Egypt	80	104.09	29	232	−58	44	204	−148
1988–1998								
Ghana	129	53.26	32	150	−30	33	111	−44
Nigeria	53	151.84	29	48	−20	59	92	−51
Egypt	68	158.98	24	65	−18	41	96	−38
1998–2008								
Ghana	46	43.62	28	248	−67	66	329	−294
Nigeria	40	177.04	29	107	−48	76	215	−191
Egypt	50	198.15	20	58	−21	45	112	−58
2008–2013								
Ghana	24.65	34.03	13	85	−40	55	279	−234
Nigeria	0.75	4.61	14	14	−23	1794	1755	−3449
Egypt	2.47	14.58	11	20	−23	428	741	−1069
2013–2017								
Ghana	16.50	28.39	9	16	−9	59	99	−58
Nigeria	15.35	95.34	−70	291	−2	−844	955	−11
Egypt	17.21	104.27	0	21	−3	0	120	−20
1978–2017 (whole period)								
Ghana	460	164.67	175	2876	−93	59	197	−156
Nigeria	223	494.20	−26	3885	−89	−25	315	−189
Egypt	446	580.07	114	1153	−80	45	149	−94

a result of Ghana's GDP declining from 2008 to 2013. During these periods, growth is also influenced by affluence and population expansion. As a result of growing urbanization and industrialization, these DMC are essential for the expansion of transit infrastructure, industries, enterprises, and residential structures.

4.4 Productivity Efficiency Transition in Ghana

In this work, the DEA-based Malmquist Productivity Index (DEAM) model was used to analyze the productivity efficiency transition. Economic growth was the end result, with material intensity (DMC per mil. \$) and CO_2 intensity (CO_2 per mil.

$) being the two inputs (GDP). This research, which examines the overall technical efficacy of Ghana, Nigeria, South Africa, and Egypt from 1978 to 2017, used the input-oriented constant return scale (CRS) as a metric.

The total factor productivity change (TFPCH) and its constituent parts are shown in Table 3 for four economic growth stages: Ghana, Nigeria, South Africa, and Egypt. Table 3 shows that over the past four decades, resource productivity has continuously increased: the average annual TFPCH was 1.097, TECHCH was 1.053, EFFCH, PECH, and SECH were 1.042, 1.001, and 1.04, respectively. Additionally, the TECHCH and EFFCH indicators helped these nations' total resource productivity efficiency and eco-efficiency to rise. It is shown that material consumption has essentially reached the optimal consumption scale while taking into account the scale efficiency growth rate of 0.1% between 1978 and 2017. Two of the deconstructed parts of EFFCH are PECH and SECH. In this instance, both indexes have simultaneously contributed chronologically to improving the value of EFFCH.

1978 and 2003, all of the countries under study had comparable developments, according to TECHCH. However, with a 24% growth rate in the stage after, Nigeria became the top nation in terms of technological innovation (during 2005–2006). Egypt has excelled other nations in terms of high-tech performance even though it is an industrialized nation with a mature economy. Since then, the TECHCH's stability has gradually improved (Fig. 8b). Figure 8c–e show the EFFCH and its deconstructed components (PECH and SECH). These indicators in Ghana are essentially consistent with its TFPCH for the years 1978–1984. It shows that the TFPCH is significantly affected by the EFFCH indicator. Overall, it took several years for all economies to reach their ideal scale. Nigeria saw the highest volatility throughout this period, demonstrating that the nation was taken into consideration for its size and historical pure change. In contrast, Egypt maintains these signs; it appears that this nation is happy with what has been constructed in the past using the inherent science and synchronization. This country concentrated on embracing global science rather than changing significantly in terms of its size or efficiency.

The geometric mean for each indicator at each country is displayed in Fig. 8f as a whole. The average TFPCH value is highest in Nigeria at 14.5%, followed by Ghana at 11%, South Africa at 9.2%, and Egypt at 4.2% growth rate, in that order (Fig. 8f). However, technical effectiveness frequently varies hand-in-hand with TFPCH. Nigeria continues to lead with a TECCH growth rate of 7.6%, while other nations only expanded at a pace of about 4%. The fact that the distance to the border decreased over the entire period through the Malmquist DEA results, demonstrating that all of these countries are making substantial effort, demonstrates that the TECH index has a significant impact on Nigeria's TFPCH growth. Nigeria's EFFCH and PECH indexes have barely increased when compared to the other countries analyzed. Nigeria's actions, which have significantly improved indicator performance, could be the root of the problem. Only Ghana's SECH is greater than Nigeria's, but both fast emerging nations (Nigeria and Ghana) do better on almost all metrics than developed nations (South Africa and Egypt) (Figs. 1 and 8f). The dominance of developing nations is evident, and they are catching up to developed nations quickly. Basically, this means boosting both the material's value and eco-efficiency. A further

Table 3 Malmquist index summary of 4 countries from 1978 to 2017

	TFPCH	TECHCH	EFFCH	PECH	SECH
1978–1979	1.069	1.052	1.017	0.995	1.022
1979–1980	2.392	1.076	2.224	1.5	1.483
1980–1981	0.743	1.12	0.663	0.79	0.839
1981–1982	1.235	1.104	1.118	1.024	1.092
1982–1983	1.322	1.092	1.211	1.06	1.143
1983–1984	0.903	1.03	0.876	0.954	0.919
1984–1985	1.076	1.134	0.949	0.929	1.022
1985–1986	1.075	1.056	1.018	0.987	1.031
1986–1987	1.091	1.096	0.995	0.964	1.032
1987–1988	1.124	1.058	1.063	1.044	1.018
1988–1989	1.16	1.066	1.088	1.077	1.01
1989–1990	1.002	1.044	0.96	0.95	1.01
1990–1991	1.093	1.063	1.028	0.99	1.038
1991–1992	1.097	1.016	1.08	1.015	1.065
1992–1993	1.094	1.006	1.088	1.012	1.075
1993–1994	1.073	0.974	1.101	1.03	1.069
1994–1995	1.099	1.049	1.047	0.995	1.053
1995–1996	1.054	1.041	1.013	0.971	1.043
1996–1997	1.03	1.028	1.002	0.955	1.05
1997–1998	1.146	1.021	1.123	1.045	1.075
1998–1999	1.069	0.974	1.098	1.029	1.068
1999–2000	1.052	1.033	1.018	0.991	1.027
2000–2001	1.062	1.035	1.026	0.966	1.062
2001–2002	1.065	1.002	1.064	0.986	1.078
2002–2003	1.064	1.022	1.042	0.974	1.069
2003–2004	1.081	1.032	1.048	0.993	1.055
2004–2005	1.144	1.064	1.075	1.01	1.064
2005–2006	1.112	1.085	1.025	1.006	1.019
2006–2007	1.11	1.088	1.02	1.001	1.019
2007–2008	1.056	1.063	0.993	0.972	1.021
2008–2009	1.049	1.041	1.008	0.975	1.034
2009–2010	1.073	1.06	1.012	1.005	1.007
2010–2011	1.067	1.037	1.029	1.01	1.018
2011–2012	1.095	1.056	1.036	1.031	1.005
2012–2013	1.079	1.072	1.006	1	1.006
2013–2014	1.061	1.073	0.989	0.973	1.016
2014–2015	1.063	1.076	0.988	0.975	1.014

(continued)

Table 3 (continued)

	TFPCH	TECHCH	EFFCH	PECH	SECH
2015–2016	1.069	1.054	1.014	1.001	1.014
2016–2017	1.098	1.087	1.01	1.002	1.008
Average	1.097	1.053	1.042	1.001	1.04
	9.7	5.3	4.2	0.1	4

factor contributing to the rising productivity indices in all nations is high technology. Furthermore, it seems that Nigeria is always catching up to technological advancements in industrialized nations. Ghana's economy, which is rapidly developing, is also going through important changes year after year. The nation has followed a relative trend in relation to developed nations like Egypt and South Africa. It forecasts that Ghana will be able to address its current problems in the near future using reasonable measures that will be advantageous to the nation in the long run.

5 Implications for Peace Through Sustainable Resource Use and Consumption

The findings presented in this research shed light on the intricate interplay between material flows, environmental impact, and the potential implications for peace, particularly in the context of Ghana's resource consumption patterns. The reciprocal relationship observed between material use and environmental consequences underscores the urgent need for a more sustainable approach to resource utilization. This discussion section delves into the implications of these findings for peace and stability, both within Ghana and on a broader global scale.

The impending resource deficit resulting from the disparity between Ghana's imports and exports raises concerns about the potential for conflicts and instability. Scarce resources can become a catalyst for competition, tensions, and even violence, as nations and communities vie for access to vital materials. By adopting a sustainable approach to resource consumption, Ghana can mitigate the risk of resource-driven conflicts and contribute to a more peaceful regional environment. Also, the dynamic resource transformation experienced by Ghana, fueled by affluence and technological advancements, provides an opportunity to align economic growth with sustainable development goals. Sustainable development, which encompasses social, economic, and environmental considerations, has been recognized as a key driver of peace. Ghana's efforts to enhance resource efficiency and reduce material intensity can contribute to the creation of a stable socio-economic foundation, fostering peaceful conditions within the country. However, the experiences of South Africa and Egypt, which have made significant strides in terms of sustainable development, offer valuable lessons for Ghana. By emulating successful strategies from these countries, Ghana can accelerate its transition towards more sustainable resource use

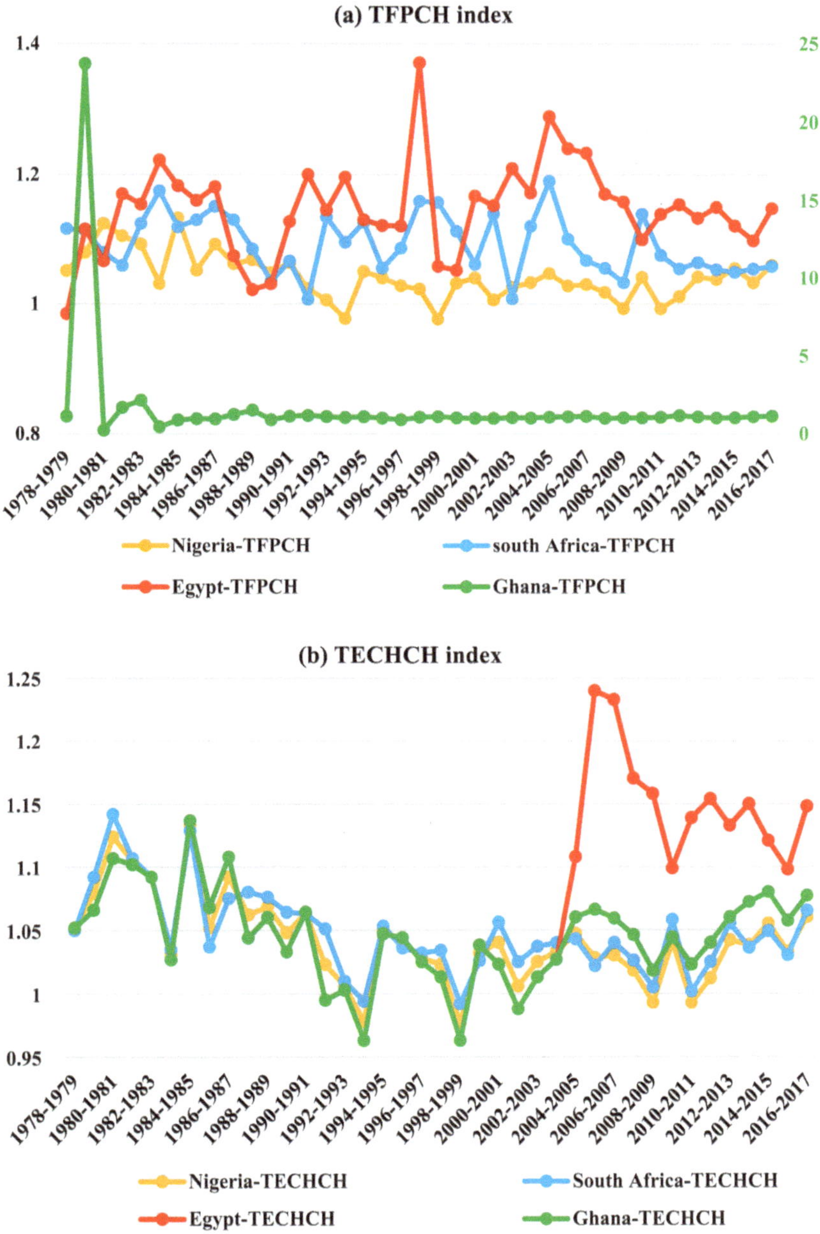

Fig. 8 Ghana, Nigeria, South Africa, and Egypt from 1978 to 2017, the Malmquist productivity index was as follows: for each nation, the average Malmquist index summarizes the total factor productivity change (TFPCH), technology change (TECHCH), efficiency change (EFCH), scale efficiency change (SECH), pure efficiency change (PECH), and efficiency change (EFCH). Note that Ghana's indicator is shown on the second axis. Since then, comprehensive outcomes have been made accessible at TFPCH represented it in (**a**). TFPCH trends varied from 0.98 to 1.4 in Egypt, South Africa, and Nigeria. The TFPCH growth rate from 1978 to 1984 was inconsistently run, with the exception of Ghana

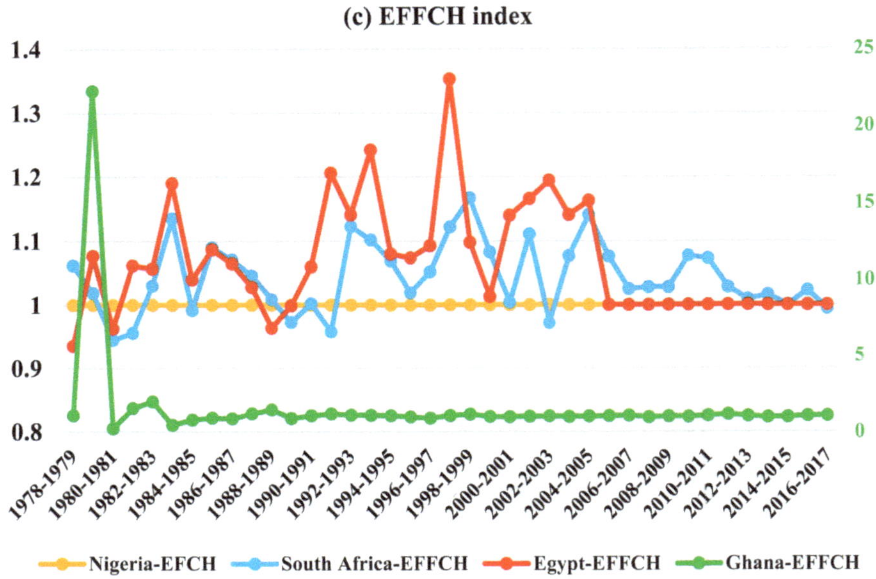

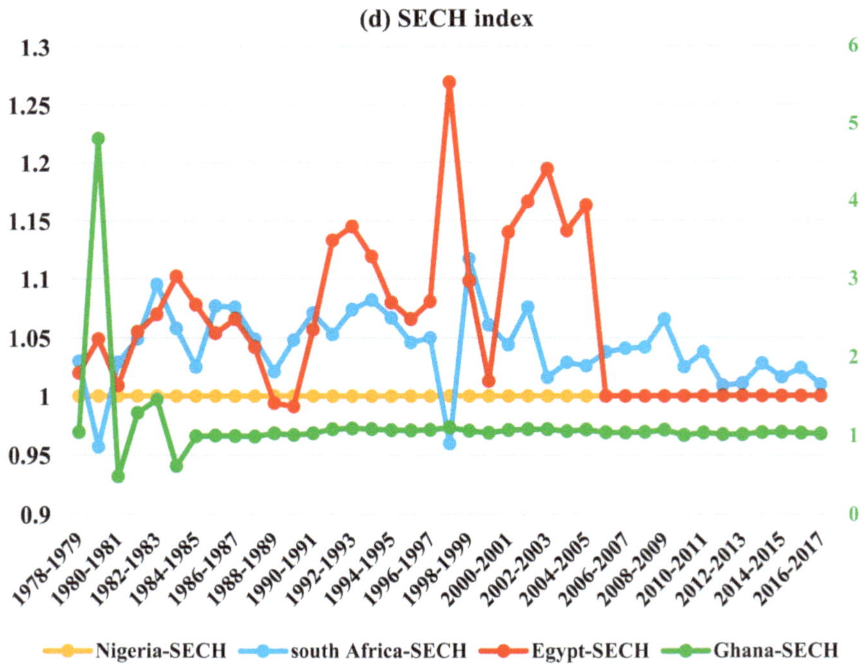

Fig. 8 (continued)

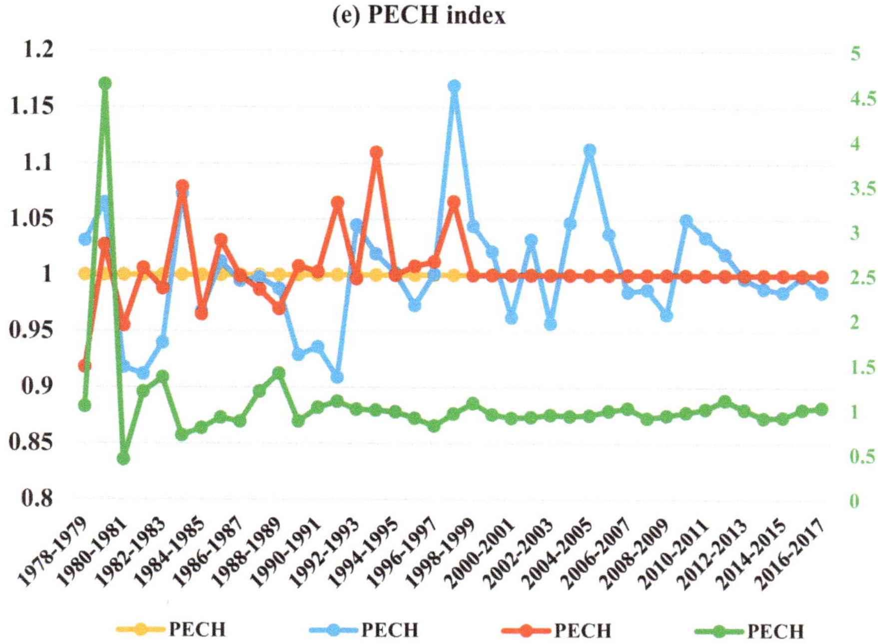

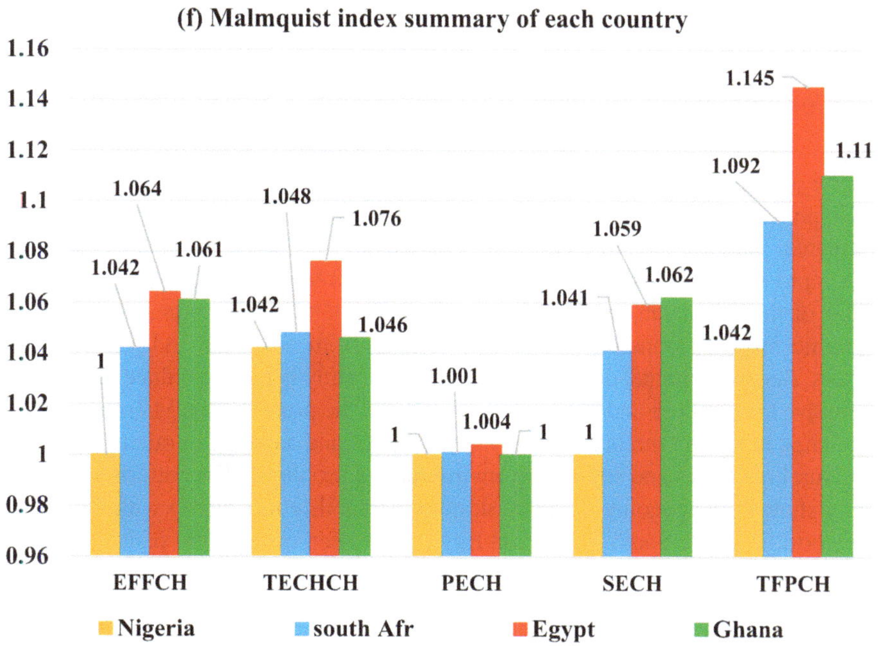

Fig. 8 (continued)

and consumption patterns. Collaborative learning and knowledge-sharing with these nations can pave the way for regional cooperation and stability.

The findings also emphasize the importance of robust policy and legislative frameworks aimed at achieving sustainable resource use. Ghana should leverage its research findings to inform the development, enactment, and enforcement of policies that promote reduced material intensity and enhanced resource productivity. Such policies can contribute to long-term peace by minimizing environmental degradation and fostering responsible resource management. In addition to this, the Malmquist-DEA efficiency index has emerged as a valuable tool for enhancing the value of natural resources while addressing environmental challenges. Ghana's utilization of this index highlights its commitment to achieving economic transformation in tandem with environmental sustainability. This approach not only bolsters the economy but also lays the foundation for peaceful coexistence by reducing resource-related. In conclusion, the implications drawn from the research findings underscore the pivotal role of sustainable resource use and consumption in fostering peace and stability. Ghana's proactive measures to align its policies with successful sustainable development models, reduce resource intensity, and enhance efficiency can contribute to regional and global peace efforts. By addressing the environmental impact of material consumption and prioritizing responsible resource management, Ghana can not only secure its own socio-economic future but also contribute positively to the global quest for sustainable peace.

6 Conclusion

Over a four-decade period, Ghana's material flows and related indicators have been on display, demonstrating the reciprocal relationship between material use and environmental impact. The input and output indicators from the MFA analysis have led to a rapid rise in raw material consumption. A resource deficit is imminent as a result of the mismatch in Ghana's import and export. Furthermore, the country's dynamic resource transformation has benefited from affluence and technological factors. Recent technical development has significantly helped to reduce the issue of excessive material use and increase the benefits of the product. The Malmquist-DEA efficiency index, on the other hand, has helped Ghana advance in raising the value of natural resources and stabilizing environmental problems. The pattern has historically shown the economy's efforts, despite its inability to keep up with the rest of the world. Because South Africa and Egypt have been on the right path in terms of sustainable development, Ghana must also continue to follow their examples. In addition, before being put into effect, laws and programs aiming to achieve sustainable use of natural resources through decreased material intensity and enhanced resource productivity should be examined, examined, and forecasted. Ghana should as soon as possible modify and tighten its development policies while reducing its dependency on reserve resources. By addressing the environmental impact of material consumption and prioritizing responsible resource management, Ghana can not only secure its own socio-economic future but also contribute positively to the global quest for sustainable peace.

7 Patent

Nomenclature

MFA	Material flow analysis
DEA	Data envelopment analysis
M	Malmquist data envelopment analysis
DMC	Domestic material consumption
CRS	Constant returns to scale
PTB	Physical trade balance
DE	Domestic extraction
MP	Material productivity
MI	Material intensity
IPAT	Impact of human activity on the environment
P	Population
A T	Affluence Technology
LMDI	Log-Mean Divisia Index
EKC	Environmental Kuznets Curve

References

Afsharian M, Ahn H (2015) The overall Malmquist index: a new approach for measuring productivity changes over time. Ann Oper Res 226:1–27

Ang BW, Liu FL (2001) A new energy decomposition method: perfect in decomposition and consistent in aggregation. Energy 26:537–548

Balat M (2004) Material flow analysis and its status in Europe countries. Energy Explor Exploit 22:441–455

Barker, Üngör M (2019) Vietnam: the next Asian tiger? North Am J Econ Financ 47:96–118

Bass S et al (2009) Integrating environment and development in Viet Nam: achievements, challenges, and next steps. Idea, p 51

Calvo G, Valero A, Valero A (2016) Material flow analysis for Europe: an exergoecological approach. Ecol Ind 60:603–610

Chang CW, Wu KS, Chang BG (2019) Productivity change and decomposition in Taiwan bakery enterprise-evidence from 85°C Company. Sustainability 11:7077

Charnes A, Cooper WW (1984) Preface to topics in data envelopment analysis. Ann Oper Res 2:59–94

Chika M, Aoki-Suzuki C, Bengtsson M, Hotta Y (2012) International comparison and suggestions for capacity development in industrializing countries: policy application of economy-wide material flow accounting. J Ind Ecol 16:467–480
De Castro Camioto F, Do Nascimento Rebelatto DA, Rocha RT (2016) Energy efficiency analysis of BRICS countries: a study using data envelopment analysis. Gestao e Producao 23:192–203
Dong L et al (2017) Material flows and resource productivity in China, South Korea and Japan from 1970 to 2008: a transitional perspective. J Clean Prod 141:1164–1177
Eurostat (2001) Economy-wide material flow accounts
Eurostat (2013) Economy-wide material flow accounting (EW-MFA): compilation guide 2013. Luxembourg: Statistical Office of the European Communities, 2013
Eurostat (2017) Economy-wide material flow accounting (EW-MFA): compilation guide 2017. Luxembourg: Statistical Office of the European Communities, 2007
Giljum S, Dittrich M, Lieber M, Lutter S (2014a) Sustainable apparel materials. Resources 3:319–339
Giljum S et al (2014b) Global patterns of material flows and their socio-economic and environmental implications: a MFA study on all countries world-wide from 1980 to 2009. Resources 3:319–339
Hammer M, Hubacek K (2003) Material flows and economic development material flow analysis of the Hungarian economy. Int Inst Appl Syst Anal
Huong TT, Shah IH, Park HS (2021) Decarbonization of Vietnam's economy: decomposing the drivers for a low-carbon growth. Environ Sci Pollut Res 25:518–529
Konrad Adenauer Stiftung. Viet Nam Quarterly Macroeconomic Report 2017, pp 1–40
Kósi K, Torma A (2005) Tracing material flows on industrial sites. Period Polytech Soc Manag Sci 13:133–150
Li J, Ma J, Wei W (2020) Analysis and evaluation of the regional characteristics of carbon emission efficiency for China. Sustainability 12:3138
Liu W, Tian J, Chen L, Lu W, Gao Y (2015) Environmental performance analysis of eco-industrial parks in China: a data envelopment analysis approach. J Ind Ecol 19:1070–1081
Long R, Ouyang H, Guo H (2020) Super-slack-based measuring data envelopment analysis on the spatial–temporal patterns of logistics ecological efficiency using global Malmquist index model. Environ Technol Innov 18:100770
Lozowicka A (2020) Evaluation of the efficiency of sustainable development policy implementation in selected EU member states using DEA. The ecological dimension. Sustainability 12:435
Nakicenovic N, Swart R (2000) Intergovernmental panel on climate change: emissions scenarios. IPCC, Geneva, Switzerland, Cambridge
Pivnenko K, Laner D, Astrup TF (2016) Material cycles and chemicals: dynamic material flow analysis of contaminants in paper recycling. Environ Sci Technol 50:12302–12311
Report C (2018) Vietnam, pp1–19
Roh C-Y, Park C, Moon MJ (2011) Economic performances of U.S. non-profit hospitals using the Malmquist productivity change index. J Manag Market Res 8:1–16
Schandl H, Turner GM (2009) The dematerialization potential of the Australian economy. J Ind Ecol 13:863–880
Schandl H, West J (2012) Material flows and material productivity in China, Australia, and Japan. J Ind Ecol 16:352–364
Shah IH, Park HS (2020) Chronological change of resource metabolism and decarbonization patterns in Pakistan: perspectives from a typical developing country. J Ind Ecol 25:144–161
Shah IH, Dong L, Park HS (2020) Characterization of resource consumption and efficiency trends in Bangladesh, India and Pakistan: Economy-wide biotic and abiotic material flow accounting from 1978 to 2017. J Clean Prod 250:119–136
The World Bank (2017) World Development Indicators. National accounts data, and OECD national accounts data files
UNDP. Fossil fuel fiscal policies and greenhouse gas emissions in Viet Nam
Update AN, Opment RED (2012) Taking Stock an Update on Vietnam' S Recent Economic Devel-an Update on Vietnam S Recent Economic 2012

Wang H, Hashimoto S, Moriguchi Y, Yue Q, Lu Z (2012) Resource use in growing China: past trends, influence factors, and future demand. J Ind Ecol 16:481–492

Wang S, Zhu X, Song D, Wen Z, Chen B, Feng K (2019) Drivers of CO_2 emissions from power generation in China based on modified structural decomposition analysis. J Clean Prod 220:1143–1155

Wiedenhofer D, Fishman T, Lauk C, Haas W, Krausmann F (2019) Integrating material stock dynamics into economy-wide material flow accounting: concepts, modelling, and global application for 1900–2050. Ecol Econ 156:122–133

World Bank (2019) World Development Indicators. National accounts data, and OECD national accounts data files

York R, Rosa EA, Dietz T (2003) STIRPAT, IPAT and impact: analytic tools for unpacking the driving forces of environmental impacts. Ecol Econ 46:351–365

Zhang Y, Yu Z, Zhang J (2021) Analysis of carbon emission performance and regional differences in China's eight economic regions: based on the super-efficiency SBM model and the Theil index. Plos One 16:e025099

Assessment of the Brokerage System in Taiwan

Anna Joceline Dizon Ituriaga

Abstract This paper focuses on the promotion and protection of migrant workers' welfare and demonstrates how the Taiwanese government failed to recognize the importance of safeguarding its migrant workers. Migrant workers are burdened by the brokerage system, which charges them a huge amount of money. Previous research has concentrated on the overall nature of migrant worker circumstances and the exploitative practices of the brokerage system; however, few studies have looked into government-to-government recruiting (G2G) as a response to the current brokerage system that exploits migrant workers. To fill this gap, this paper assesses the current brokerage system of Taiwan and explores South Korea's Employment Permit System as a valuable source of policy learning for entirely eliminating the private brokerage system in Taiwan. Despite the obvious benefits that migrant workers bring to Taiwan, I argue that Taiwan hasn't done enough to protect them. Taiwan should share the global vision of a migrant recruitment system that protects workers and meets local economic needs without putting migrant workers at risk of being highly exploited. This study draws on semi-structured interviews and builds on existing literature on migration and private brokerage agencies. This paper comes to the conclusion that the first step for Taiwan is to dispel the misconception that businesses and migrant workers are locked in an unwinnable battle over recruitment costs. The second step toward eliminating the private brokerage system is to learn about other systems, like the Employment Permit System in South Korea.

Keywords Taiwan · Migrant workers · Brokerage system · Migration

A. J. D. Ituriaga (✉)
National Chengchi University, Taipei, Taiwan
e-mail: ituriaga.anna@gmail.com

A. Sharifi et al. (eds.), *Bridging Peace and Sustainability Amidst Global Transformations*, World Sustainability Series,
https://doi.org/10.1007/978-981-99-7572-3_11

1 Introduction

Taiwan's economy flourished in the 1970s and 1980s, placing the island among the ranks of newly industrialized states alongside other East Asian countries. Many small and medium-sized businesses (SMEs) in Taiwan were forced to relocate their factories to other countries in the late 1980s due to structural changes like the sudden appreciation of the Taiwanese currency and increases in the cost of labor, particularly to the west to the Pearl River Delta in southern China. However, businesses that relied heavily on human labor to create their products stayed put (Deng et al. 2020). Because of a lack of available semi-skilled and unskilled workers, many Taiwanese companies had trouble filling positions at lower skill levels. As the birthrate has dropped over the past few decades, the number of available workers in rural areas has shrunk, Taiwan's industrial and occupational landscape has shifted, and a serious labor shortage has emerged. There has been a dramatic shift in the aspirations and career preferences of its population. With rapid economic development and improved access to education, the younger generation increasingly seeks higher-skilled and knowledge-based employment opportunities. This change in mindset has resulted in a decreasing interest among Taiwanese individuals to work in low-skilled jobs. Rather than being perceived as desirable career paths, these positions are often seen as stepping stones or temporary solutions for individuals seeking better prospects. In many societies, including Taiwan, there exists a certain stigma associated with low-skilled jobs. These positions are often perceived as less prestigious or less socially desirable compared to jobs that require higher education or specialized skills. This perception can deter Taiwanese individuals from considering low-skilled work, as they may prioritize societal status and self-image over immediate employment opportunities. These factors collectively contribute to a reluctance among Taiwanese to work in low-skilled jobs, necessitating the hiring of migrant workers to fill these positions in Taiwan labor market.

In 1989, the government of Taiwan issued its first labor migration strategy in response to this issue. The government of Taiwan has begun recruiting workers from its surrounding countries as part of a program called "Response Measure on Human Resources Demand for Fourteen Major Construction Projects" (Deng et al. 2020). To keep Taiwan's economy competitive internationally, this action led to a rise in the number of migrant workers from Southeast Asia. However, under this legislation, the newly recruited employees are no longer considered "foreign workers" but "complementary labor forces." Reports of labor exploitation and an increase in the number of undocumented migrants prompted the government to enact new legislation on labor migration named the "Employment Service Act," it legalized the import of foreign low-skilled employees to Taiwan and expanded the controlled importation of Southeast Asian migrant labor into the private sector (Schlund-Vials et al. 2019).

As of writing, there are a total of 728,081 migrant workers (TIGNN 2023) in Taiwan. Most of these migrant employees were met with challenging application processes while trying to apply for foreign contract employment due to a lack of knowledge about administration requirements in both their home country and the nation where they would be working. Therefore, they depend on labor recruiting firms to give them information on contract work in Taiwan and to help them apply for such jobs. Due to the high demand and poor supply of private employment agencies, some private recruiters have begun passing on the expenses of recruiting and placing migrant workers to the employees themselves (Deng et al. 2020).

This private brokerage system places a heavy burden on migrant labor, and Taiwan has to be aware of this. This article claims that, despite Taiwan's reputation as a liberal democracy and supporter of human rights, the government has failed to do two things: (1) safeguard its migrant workers; and (2) adhere to the International Labor Organization's basic principles in fair recruitment that respect, defend, and fulfill globally recognized human rights, especially those contained in international labor standards. This study begins by outlining the conditions underlying Taiwan's recruitment of migrant workers and then proceeds to examine the difficulties encountered by foreign workers in Taiwan. Then it went on to discuss the recruiting fees that Taiwan allows and how they benefit Taiwanese firms at the cost of migrant workers. Finally, this study investigates the viability of South Korea's employment permit system to be utilized as a model by Taiwan regarding recruiting its migrant workers. I argue that the government should make protecting the rights of migrant workers a major priority. After all, safeguarding the rights of the governed is fundamental to a functioning democracy. The Tsai government must address the rising concerns of migrant workers by modifying the Employment Service Act in Taiwan and replacing the scheme with a program that is in line with the guidelines of the International Labor Organization for the open and honest recruitment of migrant employees.

Utilizing existing literature as well as interviews with Filipino migrant workers in Taiwan, this study will explore the current challenges posed by the brokerage system. To better understand the challenges that migrant workers encounter, I conducted interviews with migrant workers to gather their views on the difficulties they confront throughout the migration process, discrimination, and working conditions. The data is based on my conducted interviews in the Zhongshan area where many Filipino migrant workers stay every weekend during their day off and on my literature review.

In the next section of this article, I discuss the overview of migrant workers in Taiwan and explain the main reasons for the challenges faced by migrant workers. Next, I discuss the Employment Permit System of South Korea. Furthermore, the article will look into the viability of adopting a similar policy to South Korea's Employment Permit System to lessen placement costs that burden migrant workers. The last section offers some summary remarks.

This study draws on semi-structured interviews and builds on existing literature on migration and private brokerage agencies. This article draws on 10 semi-structured interviews conducted in the Zhongshan district in Taipei.

2 Overview of Migrant Workers and the Challenges They Face

The labor migration procedure and the high upfront costs in Taiwan have created a two-tier structure in the recruitment of migrant workers. Those in the highest position are professionals and highly skilled laborers. It is normal practice for businesses to include the cost of hiring new staff in their overall budget for human resources. A distinct set of criteria applies to the hiring of low-skilled individuals, who are the least able to pay advanced fees owing to their lower salaries (Ling 2020).

The money might be used for anything from plane fare to local transportation to a health check to a visa to the cost of training and certification exams or the fees charged by a local employment agency. Almost the whole first year on the job is usually spent by the employee paying off the debt. In the study of Le (2019), to sign a three-year employment contract, the brokerage system charges migrant workers anything from NT$60,000 (US$2,000) to NT$200,000 (US$7,000), depending on the worker's place of origin and the nature of the job. In addition to the initial placement cost, labor brokers charge ongoing monthly fees of NT$1,500 (US$50) to NT$1,800 (US$60) for the term of the contract. While these charges are technically for service, they are frequently grossly out of proportion to what was spent. I conducted interviews among Overseas Filipino Workers (OFWs) and according to them, while the exact cost to work in Taiwan differs from brokerage firm to brokerage firm, Filipinos may expect to pay at least 60,000 Philippine pesos (about USD1, 000) to a recruitment agency. Additional charges are not included. Many Filipino families find themselves in difficult economic positions as a result of these exorbitant costs. As a consequence, some of them are compelled to borrow money from other family members or financial institutions to meet the expense of recruitment costs.

In 2016, the Department of Information Services announced changes to enhance the lives of migrant workers. They eliminated the requirement that employees leave Taiwan for at least one day following a three-year contract, and they revised the Act to prohibit the charge of extra fees, promising to shut down and penalize organizations that violate the law. The law was also changed by the relevant government agency to make it illegal for recruitment agencies to charge extra fees. Migrant workers are often exploited because they have to pay high fees, which can lead to debt bondage and poor working conditions, hence this step was implemented to tackle this problem. The government has pledged to take decisive action against any groups discovered to be in violation of this law, with the explicit purpose of either closing them down or penalizing them severely. However, it did not stop these actors from padding their pockets on the backs of the migrant workers. This unfortunate reality highlighted the ongoing challenges faced by migrant workers in Taiwan and the need for continued vigilance and enforcement to protect their rights. In a complex, transnational recruiting and placement system, it is difficult to ensure compliance with these regulations. It was reported that foreign workers in Taiwan continue to be at risk of "exploitation" and "severe financial burdens" in the 2018 United States Country Report on Human Rights Practices (Aspinwall 2019).

2.1 *Taiwan's Failure of Adhering to the International Labour Organization's Guidelines*

According to the International Labor Organization (2018), there were 164 million migrant workers in 2018, a nine percent rise from 2013. Worker vulnerability to debt, bondage, and forced labor is heightened when it comes to recruitment-related abuses, such as the collection of recruiting fees and associated charges. Those who take out loans to work may find themselves in a bind if they are unable to leave low-paying, abusive jobs or if they are unable to speak out about their working conditions for fear of losing their debts or their jobs (Alsamawi et al. 2020).

A clear framework has been developed through standard-setting documents from UN entities like the International Labour Organization and the International Organization for Migration for migrant recruiting systems to evolve toward the norm of employers paying the whole cost of recruitment. Because of the high expense of recruiting, international conventions against slavery explicitly forbid debt bondage among migrant laborers (Ling 2020).

In March 2019, the International Labour Organization Governing Body accepted definitions of recruiting fees and associated expenditures, reflecting the organization's recognition of the dangers to employees. The International Labor Organization (ILO) states that the definitions "represent a crucial step forward in giving clarity on the nature and features of recruiting fees and associated charges and are designed to promote the formulation, monitoring, implementation, and enforcement of laws, policies, and measures aimed at the protection of workers' rights, respecting the principle that employees must not be made to pay for access to employment". The International Labor Organization (ILO) has developed the "General Principles and Operational Guidelines for Fair Recruitment" to establish a framework that ensures employees are not burdened with any fees or expenditures for their recruitment. The first step in implementing this concept involves defining the term "recruitment". The scope of this definition encompasses all employees, regardless of whether they are hired locally or from another country, through intermediaries or directly by employers, including temporary staffing agencies. To provide further clarity on the matter, the ILO has provided a definition of "recruitment fees and associated charges that employees should not pay," which is explored in this guidance.

The International Labor Organization recommends that no recruitment-related costs or fees be charged to the employee. This rule was established to protect the rights of workers, especially those in precarious situations, from exploitation. The International Labor Organization (ILO) has made it clear that workers are not expected to pay for their own recruitment in an effort to eradicate debt bondage and forced labor. This includes not only the upfront charges, but also things like travel, medical exams, and tuition. Such accusations run counter to the ideals of equitable hiring and should be dismissed out of hand.

The establishment of strong enforcement and monitoring procedures is crucial to the promotion of fair recruitment practices. Policies that are in line with the principles of the International Labor Organization (ILO) should be developed and implemented with input from governments, employers, and other interested parties. Implementing regulatory frameworks, putting in place complaint systems, and scheduling routine inspections are all possible steps in this direction. This will allow relevant parties to resolve any infractions and hold unfair hiring practices accountable.

2.2 *Taiwan's Tolerance of Recruitment Fees*

Most migrants' experiences applying to work as contract workers abroad were lengthy and challenging for a variety of reasons, including a lack of familiarity with administrative regulations in both the home and host countries, cultural shock, and language barriers. The broker system is heavily relied upon the role out of the guest worker program. Taiwanese companies often prefer hiring migrant workers through private recruitment agencies because it offers a streamlined and efficient recruitment process. Work permits, visas, and compliance with labor regulations are just a few of the many legal and administrative hurdles that must be overcome when hiring foreign workers. Private recruitment services help businesses and migrant employees comply with these requirements and manage the paperwork and documents required to do so. Although businesses benefit from the implementation of recruitment fees, migrant workers bear a disproportionate share of the cost. These costs include things like travel, paperwork, visa processing, and agency fees. These costs can quickly eat up the finances of anyone looking to relocate for work prospects abroad, especially those from less fortunate socioeconomic backgrounds. Consequently, the financial burden becomes an obstacle that may take years to overcome. The cost of recruitment fees hurts the finances of foreign workers and their families in many ways. The high fees reduce the amount of money that workers can send back to their home countries as earnings. This drop in remittances can slow down the economic growth of their families and communities, making it harder to fight poverty.

To make sure that Filipinos who work in Taiwan are treated well and with respect, the Philippines and Taiwan established an alternative way to hire Filipinos through MECO and TECO in 1999. In Taiwan, this program is called the Special Hiring Program (SHPT). The SHPT was made to give companies an alternative way to hire workers without the help of Taiwan manpower agencies (TMA) and Philippine recruitment agencies (PRA). It's also aimed to create an open structure that won't exploit Filipinos in Taiwan (MECO). Private recruitment agencies are still preferred by Taiwanese enterprises despite the existence of the SHPT. This is because these agencies offer a variety of benefits to their clients, including specialized knowledge, a larger talent pool, faster procedures, reduced risk, and an extensive range of support services. Many businesses in Taiwan choose to cooperate with these private recruitment firms because of the ease, speed, and decreased paperwork involved in finding and recruiting foreign personnel. The corporation saves money by not having to pay the recruitment costs if workers are required to do so out of their own pockets, but this comes at the price of the employee who must spend weeks, months, or even years paying off the debt. Because it serves just the interests of the employer rather than those of the company's customers or clients, this benefit should not be provided to the employer. Exploiting migrant labor may provide significant, long-term cost reductions, perhaps in the hundreds of thousands of dollars each year. Due to the temporary nature of guest worker visa programs, businesses that regularly hire significant numbers of foreign nationals must do so on a rolling basis. For as long as there is no financial punishment for externalizing these costs via illegal and unethical deals with recruitment agencies and government officials, debt bondage will persist at a high rate (Verite 2019). This institutionalization of guest worker programs was first launched in the US, Switzerland, and Germany in the 1940s and 1950s. It is common for guest worker programs to last longer and grow larger than originally planned, creating structural links in the labor market between businesses and isolated clusters of jobs filled by foreign workers who are not integrated into the local economy. Employers become more economically and politically dependent on the programs as a result, as they "make investment decisions that anticipate migrants will continue to be accessible" (Ruhs 2002).

To combat human trafficking, U.S. Ambassador John Cotton Richmond requested on September 1, 2020, that Taiwan prohibit the collecting of recruiting fees, service fees, or deposits from migrant workers (AIT 2020). The government of Taiwan is aware of the private brokerage system's exploitation, yet it does little to stop it. In response to calls from migrant workers for an end to the broker system, the government released a statement saying it recognizes the importance of this problem but that the free market also plays a role since firms are allowed to choose their hiring practices. Taiwan continues to insist that companies hiring Indonesian migrant labor bear the full burden of the associated recruiting expenditures. It was in reaction to Indonesia's demand that their future employers in Taiwan pay eleven different payments associated with their Indonesian employees' departure to Taiwan. These costs include flight, passport, and visa fees (Chang 2009). Little can be done to protect migrant workers' rights until issues of unclear policies, limited resources, and discriminatory practices in receiving countries are addressed.

3 South Korea's Employment Permit System

For a significant portion of the 20th century, the Republic of Korea experienced a consistent outflow of people emigrating from the country. However, in the 21st century, Korea has transformed into a desirable destination due to its rapid economic expansion, industrialization, and various shifts in the labor market and demographics. These changes have resulted in a growing demand for low-skilled workers, especially in the small and medium-sized businesses across the country. In 1987, the South Korean government modified the Labor Law, which led to higher domestic labor costs, prompting many of these smaller businesses to seek cheaper labor abroad. The government of South Korea has had to cope with an inflow of foreign workers since 1991. Businesses receiving funding from outside investors were eligible for South Korea's International Training System (ITS), which was launched in November 1991.

Industrial skill learners would take part in the program for a total of 12 months, with the option to extend their time in the course by another 6 months (Yoo, 2005). But the program's implementation was marred by several complications, including an increase in illegal immigration, prejudice toward foreign employees, and a failure to uphold migrant workers' legal protections. Soon enough, it was clear that the ITS was encouraging violations of workers' rights and other fundamental freedoms. Businesses may start treating trainees poorly in terms of pay, hours worked, and social and occupational protections. The International Training System (ITS) has been criticized for contributing to a tendency in which legal migrants become illegal migrants as a consequence of trainees escaping onerous working circumstances and employers. Furthermore, many trainees followed the route to illegality as the ITS had the paradoxical effect of undocumented migrant employees being given bigger pay than authorized ITS trainees (Park 2008). The ITS was to blame for this unintended consequence. Many ITS participants overstayed their trainee visas because of a sense of obligation to pay back the substantial sums they had loaned to private recruitment organizations in their home countries or South Korea to participate in the ITS. These debts were generated by the trainees because they paid enormous fees to private recruiting organizations to take part in the ITS. Some students had invested over USD 10,000 in their education (Park 2008).

After concluding that the ITS was riddled with basic flaws and that the whole scheme was unable to be maintained, the government of South Korea came up with the Employment Permit System in 2004. This system is the current migration plan for low-skilled temporary workers. In 2007, the EPS completely replaced the ITS in all of its functions. As a direct result of this, the EPS has developed into the industry-accepted practice for the recruitment of personnel with low levels of expertise in Korea. As soon as they are granted worker status, these persons will, in theory, be eligible for the same benefits and protections under Korean law as their Korean counterparts would after they have obtained worker status.

The Employment Permit System (EPS) serves as an example of a non-seasonal temporary labor migration system. The EPS is operated solely via bilateral memorandums of understanding (MOU) between governments, and private sector recruiters and agencies play no part in its administration. These memorandums of understanding (MOUs) coordinate the recruitment, selection, placement, protection, and work-related benefits of migrant workers bound for Korea and stipulate the duties of both the government of South Korea and the governments of the countries that are providing employees. The EPS is designed to regulate and manage the influx of low-skilled foreign workers in a more structured and organized manner. It aims to strike a balance between addressing the labor needs of various industries in South Korea and ensuring the fair treatment and welfare of the foreign workforce.

By using the Employment Permit System, foreign workers are granted the opportunity to remain in the Korean economy and continue their employment for a duration of three years (EPS). The maximum amount of time that a migrant worker is permitted to stay in Korea is four years and ten months. In addition, businesses have the option of extending their workers' contracts for an additional two years, increasing the total amount of time that a migrant worker may spend in Korea to four years and ten months. Unless their employer makes a specific request to retain them on staff, foreign employees are forced to permanently leave for their home countries at the end of that term; in this circumstance, EPS workers may re-enter Korea after a six-month absence. Having said that, they will have to begin the process of recruiting employees all over again. However, similar to the laws that regulate the limits on changing jobs, migrant employees who have gone home after their E-9 visa has expired but whose employers seek to rehire them may return to Korea after just three months away and without having to go through the application and testing procedure for new EPS hiring. This is because they do not fall under the category of new hires (Kim 2015). By implementing the EPS, the South Korean government has taken steps to improve the transparency and accountability of the employment process for low-skilled foreign workers. This system helps prevent exploitation, abuse, and unfair labor practices by providing legal protection and support to these workers.

Furthermore, the EPS serves as a framework for promoting cultural exchange and understanding between South Korea and participating countries. It facilitates the sharing of knowledge, skills, and experiences among workers from different backgrounds, contributing to a diverse and inclusive working environment. The Employment Permit System has replaced the outdated International Training System in South Korea, offering a more systematic approach to the recruitment and protection of low-skilled migrant workers. It strives to ensure their rights, while also meeting the labor demands of the country's industries.

4 How Taiwan Can Adapt South Korea's Employment Permit System

The South Korean employment permit system places an emphasis on protecting worker rights by doing away with the use of recruitment firms. This prevents exploitative tactics and overcharging from being imposed on migrant employees. Taiwan can better safeguard its migrant workers and guarantee them decent working circumstances if it adopts a system like this. Employers hire foreign labor directly through a standardized and open system. This removes the need for middlemen, who can often be exploitative to workers and who charge exorbitant fees to recruit them. By adopting a direct hiring method, like the one outlined in this model, Taiwan can save money and reduce the likelihood of worker exploitation.

Worker Protection Mechanisms: Taiwan has to strengthen its worker protection systems to ensure the safety and security of its migrant workforce. The establishment of procedures for handling complaints, enforcing labor standards, preventing pay fraud, and guaranteeing safe working conditions are all part of this. Workers should be educated on their rights and given the tools they need to stand up for them through the implementation of proper training and education programs.

Transparent Recruitment Process: Both employers and potential employees should have full visibility into the hiring process. Job descriptions, credentials, and salaries should all be clearly outlined. Being transparent aids in preventing abuse, guarantees equal treatment, and improves compatibility between open positions and applicants' skills.

International Cooperation: Taiwan can work with other countries, like South Korea, that have put in place similar systems, to share best practices, talk about their experiences, and learn from their wins and failures. Collaboration with international groups like the International Organization for Migration (IOM) can also give useful tips and advice for putting the work permit system into place and making it better.

No burden of recruitment fees: The removal of recruitment fees is highly beneficial for migrant workers for several reasons. One of the major challenges faced by migrant workers is the burden of exorbitant recruitment fees charged by agencies or intermediaries. These fees often leave workers in a state of debt bondage, trapping them in exploitative employment situations and limiting their ability to improve their economic well-being.

In the Employment Permit System, migrant workers do not incur the expense of paying recruitment fees. They are no longer forced to take on excessive debts or endure unfair repayment schemes, which significantly improves their overall financial situation. With reduced financial obligations, workers have more disposable income and resources at their disposal.

This, in turn, allows migrant workers to send a larger portion of their earnings back to their families in their home countries. Remittances play a crucial role in supporting families and communities, contributing to poverty reduction and economic development in the workers' countries of origin. By freeing up more of their earnings, migrant workers can provide greater financial support to their loved ones, helping improve their quality of life and create opportunities for their children's education or investment in businesses.

Taiwan can adapt South Korea's employment permit system to its specific context and needs. To guarantee the system successfully protects the rights of migrant workers, promotes fair labor practices, and contributes to the development and well-being of both workers and the Taiwanese community, constant monitoring, evaluation, and modifications will be required.

5 Conclusions

In this paper, I argue that, despite the clear benefits to the nation as a whole, Taiwan has failed to recognize the importance of protecting its migrant workers. Employees are often exploited by being made to cover the costs of their own recruiting. Employers may shift the cost of hiring a worker from a poor Southeast Asian country onto the shoulders of the people they hire using this way. Many of these employees are left with substantial debt and are vulnerable to modern forms of slavery, such as debt bondage, because of this.

In these accounts, migrant workers are portrayed as dishonest, poorly educated, and unable to keep a secure workplace. The descriptions above paint a picture of people whose emotional instability and lack of common sense could put them and their friends in harm's way. The paper concludes by underlining the importance of Taiwan adopting South Korea's employment permit system in order to build a more equitable and inclusive labor market. It summarizes the advantages, including worker protection, a direct hiring process, efficiency and transparency, positive international image, economic gains, international cooperation, and social integration. The adoption of this system would promote Taiwan's commitment to workers' rights, strengthen its international standing, and contribute to economic development and social progress.

References

AIT (2020) Remarks by U.S. ambassador-at-large to monitor and combat trafficking-in-Persons John Cotton Richmond. In: 2020 international workshop on combating human trafficking. https://www.ait.org.tw/remarks-by-u-s-ambassador-at-large-john-cotton-richmond-at-the-2020-international-workshop-on-combating-human-trafficking/

Alsamawi A, Bule T, Cappa C, Cook H (2020) Measuring child labour, forced labour and human trafficking in global supply chains: a global input-output approach

Aspinwall N (2019) US Human Rights Report Flags exploitation of Taiwan's migrant workers. The News Lens. https://international.thenewslens.com/article/115514#:~:text=The%20US%20report%20flagged%20Taiwan's,is%20as%20large%20as%20160%2C000

Chang DT (2009) What can Taiwan and the United States learn from each other's guest worker programs? J Work Rights 14(1):3–26. https://doi.org/10.2190/WR.14.1.b

Deng J-B, Wahyuni HI, Yulianto VI (2020) Labor migration from Southeast Asia to Taiwan: issues, public responses and future development. Asian Educ Dev Stud 10(1):69–81. https://doi.org/10.1108/aeds-02-2019-0043

Kim M (2015) The Republic of Korea's employment permit system (EPS): background and rapid assessment

Le L (2019) How greedy brokers force thousands of Vietnamese workers to seek illegal work in Taiwan. South China Morning Post. https://www.scmp.com/week-asia/article/3210584/southeast-asian-activists-urge-asean-help-rescue-scam-trafficking-victims-chinese-gangs

Ling B (2020) Taiwan must protect labor migrants. Taipei Times. https://www.taipeitimes.com/News/editorials/archives/2020/11/07/2003746500

MECO. Special Hiring Program for Taiwan. https://www.meco.org.tw/labor-and-atn/special-hiring-program-for-taiwan

Organization IL (2018) New ILO figures show 164 million people are migrant workers. https://www.ilo.org/global/about-the-ilo/newsroom/news/WCMS_652106/lang--en/index.htm

Park Y-B (2008) Admission of foreign workers as trainees in Korea (9789221208426). Bangkok

Ruhs M (2002) Temporary Foreign Worker Programmes: policies, Adverse Consequences, and the Need to Make Them Work. Center for Comparative Immigration Studies

Schlund-Vials C, Beauregard G, Lee H-C (2019) The subject(s) of human rights: crises, violations, and Asian/American critique. Temple University Press, Philadelphia

TIGNN (2023) The number of migrant workers of Taiwan increased at record high with the open of the border. https://news.immigration.gov.tw/NewsSection/Detail/309f1443-1a40-4ffd-b298-3f55be3e4984?category=1&lang=EN

Verite (2019) Remediation and elimination of recruitment costs charged to migrant workers. https://verite.org/remediation-of-recruitment-costs/
Yoo K-S (2005) Foreign Workers in the Republic of Korea. Korean Labor Institute, the Republic of Korea

The Impact of Migration Acculturation Strategies on Self-continuity: A Case Study of Thai Skilled Workers in Japan

Parkpoom Kuanvinit

Abstract Due to aging population problem, international migrant workers play increasingly significant roles in the Japanese labour market. While some studies explored existing social challenges during the acculturation process, the coping mechanisms and the sense of self-continuity among the migrants in Japan are largely understudied. This study intends to achieve three objectives through analysing narratives depicted by the life-line drawing method from 12 highly skilled Thai migrant workers. The first is to identify the dominant acculturation strategies that Thai workers apply in Japanese society. The second is to explore the dynamic of self-continuity among them. The third is to uncover the correlation between acculturation strategies and the sense of self-continuity. The analysis leads to three major findings. First, the integrative approach prevails as an acculturation strategy. Second, the sense of self-continuity among Thai participants is high and consistent, even though they have been pressured to behave according to Japanese norms and values. Finally, there is the positive correlation between integration as an acculturation approach and self-continuity. In contrast to previous studies which emphasize the role of social structure in the pattern of adaptation, this research sheds light on the ability of individuals to enable their own acculturation strategies regardless of existing social norms. The findings of this research hold implications for future identity research to take into consideration the cultural and anthropological perspectives. The bottom-up approach of this study is also expected to provide insight for policymakers to design and implement proper multicultural coexistence policies for migrant workers in Japan.

Keywords Acculturation strategies · Self-continuity · Migrants in Japan · Highly skilled Thai migrant workers · Social norm

---

P. Kuanvinit (✉)
Graduate School of Humanities and Social Sciences, Hiroshima University, Hiroshima, Japan
e-mail: parkpoom.ku@gmail.com

A. Sharifi et al. (eds.), *Bridging Peace and Sustainability Amidst Global Transformations*, World Sustainability Series,
https://doi.org/10.1007/978-981-99-7572-3_12

1 Introduction

According to the Ministry of Health, Labour and Welfare of Japan (n.d.), the average life expectancy of Japanese people as of 2018 was 81.25 years for men and 87.32 years for women. The number of elders aged 65 and above has also been on the rise, which was a 28.4% increase in 2019. The number is considered among the highest in the world and is anticipated to soar to 33.3% in 2036.

What comes in parallel with the increasing number of aged people is the rising influx of migrant workers. As of 2022, up to 1.8 million foreign workers have been employed in Japan, which doubled the rate of the past decade. The increase reflects the growing demand for human resources (Ministry of Health, Labour and Welfare of Japan n.d.). Thai people are among international migrants increasingly playing significant roles in Japanese economy. Influenced by this growing trend, this research was conducted to investigate the acculturation strategies of highly skilled Thai migrant workers in Japan. Moreover, it also seeks to understand the impact of acculturation on the extent to which they maintain their original culture.

At the bilateral level, Thailand and Japan enjoyed intimate diplomatic relationship. In June 2020, 20,365 Thai people entered Japan as economic migrants. Of these, 13,505 people were identified as legal, while 6,360 were illegally employed (Office of Labour Affairs, Royal Thai Embassy n.d.). Despite a higher number of legal Thai workers in Japan, existing literature mainly explores the conditions of the illegal ones, especially sex workers (Sellek 1996; Ruenkaew 2002; Aoyama 2009). Thus, the issues of legal Thai migrant workers are understudied.

Regarding lived experienced of foreigners in Japan, a growing literature has expressed concerns about varying living difficulties. Much literature emphasizes discrimination, racism, and other forms of structural violence that foreigners are vulnerable to (Froese 2010; Kim 2011). Thus, the coping mechanisms applied by migrants to combat social challenges in Japan are considered an interesting topic. This research conceptualizes acculturation strategies as coping mechanisms for migrants which remain underexplored.

Considering ethnonationalism-centric policy (Neuliep et al. 2001), Japan represents a fascinating research site to gain insight into the acculturation strategies of migrants. Aulturation is not only conceptualized as the personalization of new cultures but also the extent to which acculturating individuals retain their pre-migration identity (Berry 2005). Therefore, this research assumes the impact of acculturation strategies on the continuation or change of migrants' original culture.

Among varying migration populations, economic migrants have become an intriguing topic of study due to a steady increase in number and economic contribution at the global scale (World Bank n.d.). Economic Commission (n.d.) defines economic migrants as individuals who willingly move to other countries for economic reasons. Economic migrants can also be classified into different levels of skills (ILO n.d.). According to ILO (n.d.), economic migrants can be classified into different levels of skills. Highly skilled ones refer to those with high education, specialized skills or knowledge for complicated tasks. In contrast, low-skilled work is associated with hand-held and physically-demanding tasks. Considering a dearth of research on the interpersonal relationships between highly skilled migrant workers and local people in Japanese context, this study was carried out to gain better insight of the lived experience of highly skilled Thai migrant workers in Japan.

In a nutshell, this research aims to understand the link between acculturation strategies and the sense of self-continuity among highly skilled Thai migrant workers in Hiroshima, Japan. The research was conducted under three objectives. The first objective is to investigate the acculturation process with an emphasis on its effect on Thai culture. Inspired by growing literature from the sociopsychological field, another fundamental objective is to understand the dynamics of self-continuity among highly skilled Thai migrant workers. As for the last objective, this article further enquires about the correlation between acculturation strategies and self-continuity. In accordance with those objectives, the main research question is, "How do acculturation strategies affect the sense of self-continuity of highly skilled Thai migrant workers in Japan?". Four sub-research questions are formulated as follows: (1) What acculturation strategies do highly skilled Thai migrant workers use to live in Japanese society?; (2) What are the reasons behind the decisions of highly skilled Thai migrant workers to opt for acculturation strategies?; (3) How do adaptation strategies change during their migration experience?; and (4) To what extent and under what conditions highly skilled Thai migrant workers can retain their self-continuity?

In terms of methodology, data was generated from life-story narratives depicted by life-line drawing by twelve Thai migrant workers aged between 25 and 34 years old in engineering, academic, and food industries. The thematic content analysis was applied as descriptive presentation of qualitative data.

This research is deemed significant for three reasons. First, the theoretical framework in which both sociopsychological and anthropological perspectives are intertwined can be fruitful for migration studies with emphasis on psychological well-being. Second, the concept of self-continuity is reconceptualized through both psychological and cultural aspects which provokes interdisciplinary discussion. Finally, the study makes recommendations for policymakers and relevant stakeholders to design and implement proper migration policies that foster multicultural and sustainable society in Japan.

Following the introduction, the theoretical framework, which features key concepts and theories, is elaborated. In the adherent section, significant literature is reviewed and circumscribed under four themes. The succeeding section elaborates the research methodology together with details of participants and research procedure. In the following section, data collection and analysis are explained. Finally, the conclusion part wraps up the overview of the entire manuscript with some recommendations for future studies and policymakers.

2 Literature Review

Relevant literature is categorized into five themes which; lived experience of international migrants, acculturation of Thai migrant in host societies, living difficulties of foreigners in Japan, self-continuity in relation to migration studies, and prevailing Japanese norms. Following the discussion on prevailing academic discussion in each theme, the research gaps to be addressed by this study are summarized.

Regarding studies on the lived experience of international migrants, most academic attention has focused on the meaning of integration (Castle 2004; Fyvie et al. 2003), the conditions or relationships necessary for integration to occur (Zetter et al. 2002), or the dynamism and fluidity of the settlement process (Ager and Strang 2008). Others express their interest in the obstacles to integration, namely perceived negative stereotypes of migrants and refugees (Steele et al. 2002; Raijman 2013), as well as the adverse outcome of failed integration on the physical and mental health of migrants (Ji et al. 2020). Even though these studies vary in their topics of studies, they bear similarities in terms of the effort to uncover the effects of social factors on the integration and migration policies in the host society.

As for acculturation of Thai migrants during sojourns, numerous studies have been conducted to explore how Thai migrants overcome hardship in host societies. The emphasis and pattern of the discussion are oriented toward the formation of Thai communities and maintenance of Thai cultural and religious identity as coping mechanisms. Therefore, scholarly attention has been centered around social contact at the intragroup level (Smutkupt and Kitiarsa 1999; Ruenkaew 2002; Tan 2018; Straiton et al. 2019; Kalapong 2023).

Based on relevant research, the living difficulties of foreigners in Japan can be classified into two main dimensions; pressure at work and social discrimination. Kim (2011) argued that the exclusionary practice against ethnic minorities manifests as a converted form of blatant discrimination against ethnic minorities in Japan. Similarly, Otsuka and Anamizu (2019) suggested that bicultural and biracial identity can be instrumentalized as a coping mechanism against discrimination in Korea and Japan. In the professional domain, employees in Japan are prone to worsening working conditions in parallel with a rising level of job requirements (Aichinger et al. 2017; Arimoto and Daizen 2012). This strand of literature highlights the existence of racism, prejudice, and discrimination against foreigners in Japan. In addition, identity as foreign workers in Japan reflects the double hardship as they are vulnerable to difficulties from both everyday and work difficulties.

Within the explanation of self-continuity as a crucial element of people's identity and mental health, self-continuity correlates with the examination of emotions. Studies of emotion within identity theory have flourished in recent years. These studies bear resemblance in the way of discussion, as they focus on the individual level of identity namely self-development, identity negotiation, brain function, and effects of traumatic experiences (Eliades et al. 2013; Leveto 2016; Trettevik and Grindal 2016; McClelland 2020). Despite a limited number of literature explores the sense of self-continuity among migrants and refugees. The prevailing trend of arguments is in relation to autobiographical meaning-making process and mechanisms to understand self-development, especially among the group of juveniles in multicultural settings (Chandler et al. 2003; Phinney 2010; Camia and Zafar 2021; Becker et al. 2018).

As this article deals with the extent to which Thai workers embrace Japanese cultures on the one hand and retain their Thai cultures on the other, it is necessary to conceptualize what has been accepted in the scholarly community as Japanese norms and values. Many articles similarly identified sense of selflessness as the prevailing norm in Japanese society. Such norms can be characterized by the rationalized message of apologies (Smith 1961), a sense of group belonging (Kamolkhantibowon 2018), and concerns about how to preserve group harmony (Takamatsu 2021). Some scholars also suggested that foreigners are likely to be integrated better if they can follow or initially possess similar norms and values (Hinenoya and Gatbonton 2000; Kamolkhantibowon 2018). More specifically, some literature focuses on collective organizational behaviors in Japanese workplace. Support for the majority and loyalty to the organization can be identified as ideal characteristics (Hinenoya and Gatbonton 2000; Nakajima 1997). Whereas these norms were perceived to preserve social order (Fontana 2020), some studies pointed out a positive correlation between these norms and the justification of bullying and discrimination in the workplace (Meek 2004; Wolf and Gan 2013).

While it is acknowledged that groups and individuals become integrated differently (Crown and Rosse 1995), there is a dearth of research regarding the varied settlement experiences of individual refugees and migrants. Seeing that the dominant trend of scholarly discussion is oriented toward group-level integration, more studies with the awareness of the roles of individual, experiential, and cultural factors in settlement experiences can extend the knowledge of the existing studies. Among an array of literature about Thai migrants in exile, the predetermined focus on the ingroup relationship generally presupposed the separation as a dominant acculturation strategy at the early stage of the studies. The in-depth analysis of the patterns and dynamics of acculturation strategies of Thai migrants in host societies remains elusive. The careful analysis of the acculturation strategies potentially brings about new dynamics and perspectives of the migratory experience of Thai people, especially in terms of coping mechanisms. Furthermore, Numerous studies on hardship of migrants in Japan concretely shed light on the low level of happiness and hardship, particularly in Japanese workplaces. However, none of the literature has extensively explored the coping mechanism of legal migrant workers against hardship in Japan. Therefore, more studies regarding adaptation strategies of economic migrants in Japan can complement previous studies.

While the concept of self-continuity has been discussed in previous literature with a focus on the individual level of emotion examination, little is known about the interactional and structural version of identity. Hence, the study of self-continuity in relation to the acculturation process can complement the existing identity research, as it takes into consideration the collective cultural identity and emphasizes the interactional feature of identity. Previous studies explained the impact of cultures on self-continuity, focusing on the roles of personal preferences toward cultures as triggers of self-continuity justification. This perspective is in contrast with the social identity theory, which emphasizes the necessity of social interaction as a major factor that develops, reshapes, contextualizes, and decontextualizes personal identity (Hogg 2003). Additionally, since related literature has explored the self-development of juveniles in multicultural context, more studies that contemplate the influences of cross-cultural encounters toward self-continuity among adults can expectedly address these shortcomings.

A sense of selflessness contextualized by support for the majority and group loyalty can be considered an ideal social value of Japan in both everyday-life and professional domains. Exaggerated version of group loyalty may result in structural violence such as social exclusion, especially in the workplace. These norms will be applied as a framework to compare with the narratives of Thai workers in Japan. The focus would be on these norms' impacts on social integration and the extent to which migrant workers can retain their home cultures.

3 Theoretical Framework

Fundamental to this study is the concept of multiculturalism. In the age of globalization, when the movements of people across states and continents have become a common phenomenon, the notion of multiculturalism has been increasingly discussed. However, as society is mobilized through the daily interaction of societal members, multiculturalism has been conceptualized beyond the *space* occupied by multinational residents. The contemporary version of multiculturalism puts more emphasis on the *process* of cross-cultural encounters toward the respect of cultural diversity (Ong et al. 1996; Berry 2005; Milner and Browitt 2013; Vink and Bauböck 2013).

Inspired by the concept of multiculturalism, Berry (2005) has developed an acculturation strategy model widely referred to in refugee and migration studies (Leong and Chou 1994; Piontkowski et al. 2002; Schwartz et al. 2008; Ward and Kus 2012). As elaborated by this framework, acculturating individuals apply four different strategies upon intergroup contacts; assimilation, integration, separation, and marginalization. Assimilation prevails when individuals abandon their original culture and personalize host cultures. As opposed to assimilation, separation is the term used in a situation when individuals reject new cultures and have a strong attachment to the original one. The integration strategy is defined in the case of individuals who holistically adopt both host and home cultures. Finally, marginalization describes situations when individuals isolate themselves because they do not feel a sense of belonging to any group (see Fig. 1).

The degree to which acculturating individuals can hold on to their original culture bears similarity to the concept of self-continuity. Erikson (1968) clarified the sense of self-continuity as the feeling of sameness in self over time. In contrast to self-continuity, self-discontinuity is mentally harmful because it indicates personality

	Reject Home culture	Accept Home culture
Accept Host culture	**Assimilation** Members highly identify with host culture but not their ethnic culture	**Integration** Members of the minority group highly identify with both the ethnic and the mainstream culture
Reject Host culture	**Marginalization** Members identify low with both cultures	**Separation** Minority members highly identify with their ethnic culture and do not interact with that of the majority

Fig. 1 Acculturation strategies. Adapted from Berry (2005, Fig. 3.3, p. 20)

disorder relating to self-harm and suicidal thoughts (Joiner and Rudd 1995). Self-continuity can be at stake during migratory experience because migrants are likely to experience some changes in their personal identity, in an attempt to gain recognition from the host society. Therefore, the correlation between acculturation strategies and self-continuity is apparent but underexplored.

As clarified under identity negotiation theory, the repetition of some behaviors provides evidence for individuals to identify who they are. However, people can also be influenced to adopt new identities, strengthen the existing ones, or erase them, most of the time to better fit into the society (Swann and Bosson 2008). Based on this theory, the self-continuity concept does not resist development of identity for social adaptation purposes. In alignment with the notions of fluid and developmental identity (Onorato and Turner 2004; Osei-Kofi 2012; Wiese et al. 2018), this study conceptualizes *change* in the form of a spectrum indicating a range of degrees from self-continuity to self-discontinuity. In other words, only the extreme degree of change indicates the complete disconnection with the past self, which positively correlates with self-discontinuity (see Fig. 2).

Acculturation is the process that emphasizes an individual's attitudinal and behavioral adjustment to another culture on the one hand and touches upon the social impact on the dynamic of acculturation on the other (Sam and Berry 2010). The concept, therefore, is posited as a holistic approach prevalent in anthropological and sociopsychological fields (Flannery et al. 2001; Cabassa 2003; Schwartz et al. 2010). Self-continuity is mostly discussed in psychological discipline (Park 2019; Camia and Zafar 2021; Sedikides et al. 2023). The theoretical framework which incorporates the two concepts makes two significant contributions. First, it takes an interdisciplinary approach of anthropological and sociopsychological studies. Second,

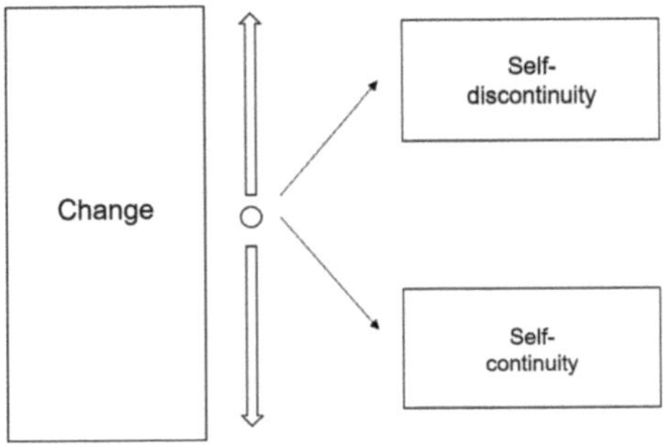

Fig. 2 The concept of change in relation to self-continuity and self-discontinuity

it reconceptualizes self-continuity through the anthropological lens to gain a better understanding of the change of continuation of cultural identity. The holistic understanding from this approach can be a useful lens to investigate acculturation and self-continuity aspects.

The following section details the reviews of related literature and existing research gaps. The literature is thematically organized around four main issues, which are the acculturation process of Thai migrants in host countries, lived experiences of international migrants, living difficulties of foreigners in Japan, and prevailing Japanese norms.

4 Methodology

4.1 Methodological Framework

In order to examine the dynamic processes of acculturation and self-continuity, research tools facilitate Thai workers to tell their own stories freely and openly, unbounded by fixed and close-ended questions. For this reason, life-story narratives depicted by life-line drawings are justified as a suitable method. As explained by Park (2019), life-line drawing is a self-reflection method that depicts the dynamics and impactful events of an individual's life course. As elaborated by Camia and Zafar (2021), self-reflection and autobiographic memory are therapeutical tools to safeguard and measure the sense of self-continuity. The use of life-line drawing, thus, can facilitate the participants to build up narratives by means of conscious past reflection. Such narratives, then, can be utilized to explore the sense of self-continuity.

Migration trajectory inevitably associates with new experiences of social and cultural encounters. Following Niedentha et al. (2018)'s postulation about migration's effects on the emotional states of acculturating individuals, the narratives on acculturation strategies are expected to emerge through the life-line drawing.

With preconceived notions of acculturation strategies, deductive coding is a method used to classify and categorize codes derived from the data and identify the acculturation strategies. On the other hand, the exploration of self-continuity is based on the inductive approach in which the codes are allowed to emerge from the narratives without predetermined themes (Elo and Kyngäs 2008; Bengtsson 2016). The correlation between acculturation strategies and the sense of self-continuity can be exposed by systematically coding and categorizing excerpts demonstrating the negative and positive perceptions of their life-courses as a consequence of the acculturation mechanisms.

4.2 Participants

Due to the COVID-19 pandemic that limited chances for physical meetings with the interview participants, twelve Thai workers voluntarily participated in interviews online using the Zoom application. Among these participants, seven of them are male, and the other five are female. Their ages range from 25 to 34 years old. They have worked in Hiroshima City for at least one year. The industries range from engineering to the food industry and academia. Over 300 leading companies in different fields are based in Hiroshima (Jetro n.d.). The latest survey from Hiroshima City illustrated that 26.6% of companies (617 out of 2,318 companies) in Hiroshima employ foreign workers. More than 70% of the companies are reportedly highly satisfied with the performance of foreign employees. The number of Thai residents in Hiroshima is also on the rise, from 836 in 2015 to 1,227 in 2019 (Hiroshima City 2019).

Taking into account the roles of language proficiency in cross-cultural communication, the ability to communicate in Japanese is considered one of the significant criteria. Referring to results from Japanese-Language Proficiency Test (JLPT), the levels of participants' Japanese skills range from beginners, intermediate, and advanced levels, which are determined by the levels of N5-N4, N3, and N2-N1, respectively.

Another criterion that makes this group of participants unique is their educational background. All the participants obtained master's and doctoral courses from Japanese graduate schools. High education, in this context, makes two major contributions. First, it ensures employment opportunities for highly-skilled migrants. Second, several years spent in Japanese universities allowed participants to learn and familiarize themselves with Japanese culture. As shown in some studies about the exclusionary attitude of some Japanese people against low-skilled migrant workers due to threat perception (Neuliep et al. 2001; Morita 2015; Tian 2019), participants in this study represent fascinating research subjects thanks to their status as highly-skilled migrant workers and acquaintance with Japanese cultures which enhances the possibility of intergroup interaction. The sociodemographic attributes of all the participants are presented in Table 1.

It is also worth noting that the patterns and frequency of interaction at work vary depending on the industry. Those who work in the food industry have varying roles and responsibilities since they need to handle both administrative and laboratory tasks. Thus, the forms of interaction with Japanese colleagues also vary. Participants from the engineering industry mostly work with their personal computers in the

Table 1 Sociodemographic variables of research participants

No	Gender	Ages	Work industries	Material statuses	Periods of stay in Japan (year)	Japanese proficiency[1]
1	Female	26	Product developer in a food industry	Single	6	Intermediate
2	Female	25	Engineer in an automotive company	Single	3	Advanced
3	Female	25	Engineer in an IT company	Single	2	Beginner
4	Female	27	Product developer in a food industry	Single	1	Intermediate
5	Female	34	Research assistant at a university	Single	3	Intermediate
6	Female	27	Engineer in an automotive company	Single	5	Intermediate
7	Female	26	Engineer in an IT company	Single	3	Intermediate
8	Male	26	Engineer in an automotive company	Single	5	Intermediate
9	Male	33	Engineer in an IT company	Single	5	Intermediate
10	Male	25	Engineer in an IT company	Single	2	Intermediate
11	Male	28	Engineer in an IT company	Single	2	Intermediate
12	Male	30	Product developer in a food industry	Single	4	Intermediate

office. Thus, correspondence with Japanese colleagues is rather limited to emails, meetings, and daily conversations. A participant in the academic field works as a research assistant. Therefore, she mainly assists a Japanese professor in academic-related tasks. Even though her main contact person is her supervisor, she occasionally encounters Japanese and international colleagues. However, in her position, Japanese proficiency is not required.

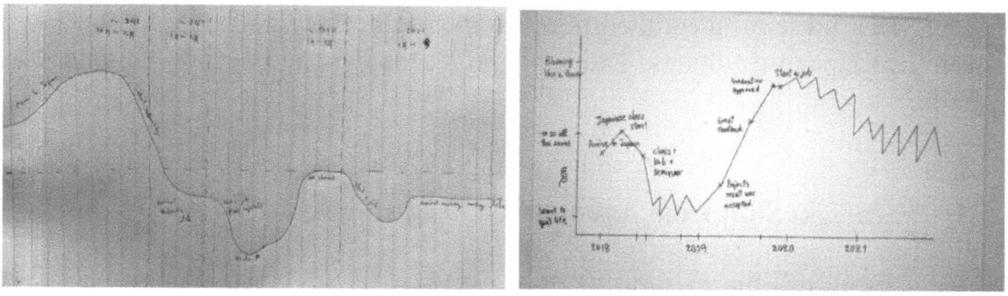

Fig. 3 Samples of life-line drawings by Thai workers

4.3 Measure and Procedure

Initially, the recruitment of the research participants was carried out through the researcher's personal connection. The snowball sampling strategy was also applied to expand the sample size. At the beginning of the recruitment process, the participants were well-informed about the research topic, objectives, and methodology. Following instructions and guidelines of the Research Ethics Review Board, Graduate School of Humanities and Social Sciences, Hiroshima University, this study was conducted with careful ethical consideration. To ensure voluntary participation, all the participants were asked to give written consent (using Google form) before the interview. For confidentiality, pseudonyms are used in case people's names are mentioned in this paper. Embracing the 'do no harm' principle, the participants were also explained their rights to skip any questions or withdraw from an interview at any time.

After the confirmation of the participants' understanding of the necessary research procedure and voluntary participation, the participants were requested to perform the life-line drawing task (see Fig. 3). Since the life-line functions as a visual plot for narrative construction, the participants were allowed to freely draw the lines without strict instructions. The only requirement is for the participants to specify major events that trigger different turning points in their life course. The participants were then asked to narrate their life stories as depicted by the life-line without or with only a few interruptions. With the participants' permission, all the interviews were recorded for further analysis.

5 Data Analysis

5.1 Narrative Analysis of Acculturation Strategies

As previously mentioned, acculturation strategies are situational and developmental, and each participant in this study has experienced different acculturation strategies. However, all the participants agreed that integration prevails as the most effective strategy. This sense of integration refers to the balance and holistic personalization of Thai and Japanese cultures. The integration progress can be classified into three patterns.

5.1.1 Paradigm Shift from Assimilation to Integration

Attempts to imitate the gestures and behaviors of Japanese people are one of the most common narratives. When imitation became natural and automatic, the sense of belonging as a member of the Japanese society erupted. Among participants applying this approach, there is also evidence of disconnection and abandonment of Thai cultural identity through the migration trajectory of some participants. According to Berry, the rejection of home culture for the sake of adjustment toward the host culture can be identified as an "assimilation" approach.

> I tried to be a member of the Japanese community, speak like them, dress like them, and go everywhere with them. I believe it would help me blend in well with society. But finally, I felt like it was just not my place. I will never be like them (Nussaba 2022).

> I felt like I didn't think much about Thainess and even regarded some Thai norms as unorganized and unpleasant. Come to think of it, I think I just tried to convince myself that I was more Japanese than Thai... I realized later that being Japanese is not that special, and I can never be one of them (Nussaba (follow-up interview) 2023).

> Before I knew it, my facial and verbal expressions became like Japanese. People often misunderstood that I am a Japanese. To be honest, I felt quite happy with that misrecognition. I was also happy hanging out with Japanese friends and hardly wished to socialize with Thai people (Phukao 2023).

The excerpts above demonstrate how participants try to behave and act out like a Japanese for the adaptation purpose. They also showed that such effort has adverse effects on Thai cultural identity. Most importantly, this approach cannot be regarded as a sustainable way to achieve positive adaptation outcomes. As illustrated by Nussaba's case (see Fig. 4), the assimilative approach over time made her feel detached from Thai culture and community. Her consecutive attempt to secure a sense of belonging among Japanese people was unfortunately in vain as she did not enjoy the sense of recognition she wished for. The failure of acculturation also affected her mental state, as visualized by a falling area of the life-line pointed out by an arrow. One of the factors that remedied her mental state was the reunion and socialization with Thai people, which explains the line going toward a rising direction. Thus, it illustrates that integration is comparatively a more acculturative approach.

5.1.2 Strategic Shift Between Assimilation and Integration

Similar to the aforementioned case, a few respondents find acceptance of both Thai and Japanese cultures useful for adjustment to Japanese culture. However,

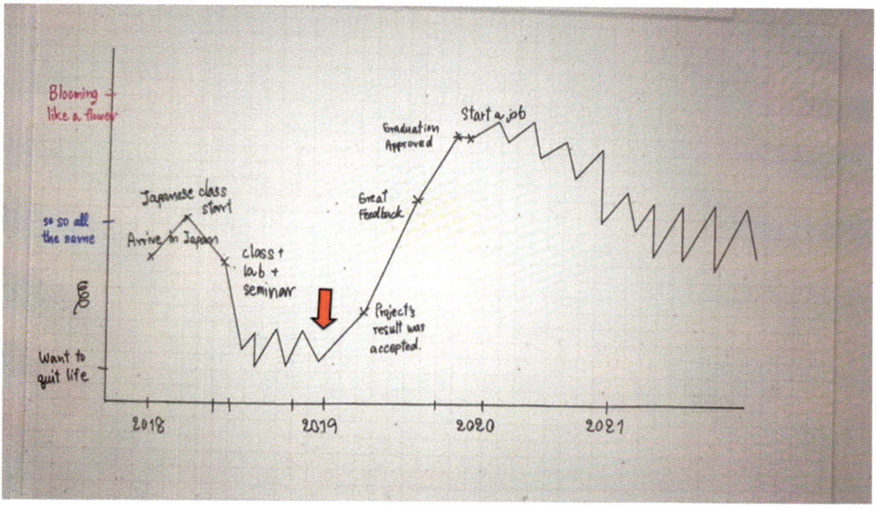

Fig. 4 Nussaba's life-line

the pattern of acculturation is slightly different. As identity can be very context-independent, many participants agreed that they can effectively participate in the society if they display Japanese culture. Since Japanese workplaces put importance on social harmony and hard work, it is essential for participants to show that they advocate for these social norms to gain social acceptance. Punctuality, nonverbal cues to show support and prioritization of group achievements over individual ones are some values that the participants strictly bear in mind in the workplace. These practices can also contradict Thai work cultures characterized by flexibility and the weak boundary between work and personal issues. Compliance with Japanese norms, on the one hand, and avoidance of Thai culture, on the other, indicate acculturation is the overall strategy in the professional domain.

Nevertheless, it is interesting that the participants regard these behaviors as a survival tool at work, meaning that they are not compulsory in other non-professional domains. To illustrate, some participants imitated and followed Japanese behaviors only at work. They even find it relaxing when they can allow themselves to be more flexible and carefree outside working hours. This group of participants also maintains their Thai culture and continues associating with the Thai community. The activation of Japanese and Thai cultures, depending on the situation, also fits the description of acculturation as the sense of acceptance of both cultures is equally vital.

> … people would exclaim "Hiss" when they change their movement…for example, after two hours of sitting position, they would say "Hiss" before standing up. Gradually, I intentionally mimicked their gesture because I believed it would help me blend in well. However, I would do so only when I am at work. After work, I enjoy socializing with Thais and hardly have that expression (RESP#8).

> I have noticed that Japanese people nod several times to show their support and agreement in the conversation. When I talk to them, I feel like I should repeatedly nod, too, to show that

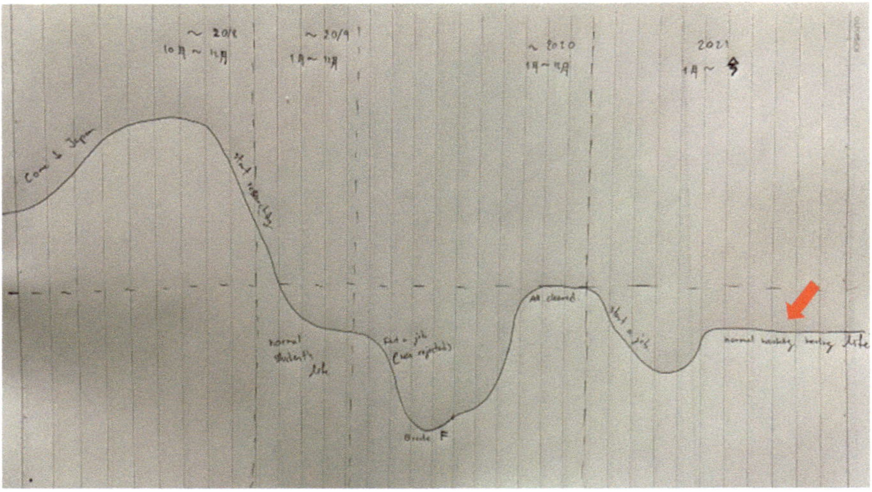

Fig. 5 Thanawat's life-line

> I really pay attention to what my colleagues are saying. Sometimes, I feel like I nod until my neck hurts, but since everybody does so, I feel obligated to do the same (RESP#2).

> I am not a very punctual person, but aware of how Japanese people are serious about time, especially at work. I can be on time when I'm at work, but I would say I become less strict when I have meetings with friends (RESP#9).

> I think I have become more independent. It's quite usual for Thai people to have lunch with friends during breaks at schools or workplaces. Sometimes, my colleagues went ahead to have lunch without me, so I was worried whether they disliked me. I realize later that it's quite common to have lunch alone, so I do the same these days (RESP#1).

The findings from this group of participants emphasize the role of social structure in coercing and determining the set of cultural values that group members are expected to follow, even at the expense of neglecting individual pre-existing cultures. Additionally, they also highlight individuals' ability to retain their original cultures, even though they are encouraged to periodically deactivate them in some situations. None of the participants relate the switch between the two approaches with the sense of identity disruption or any signs of mental disorder. As evidenced by Thanawat's life-line, the area of the lines that accompany depicting the period when he needs to shift between acculturation and integration seems stable (see Fig. 5).

5.1.3 Consistent Integration with the Perception of Japanese Culture as Self-Enrichment

Whereas the previous two patterns signify some changes between different strategies, some participants manage to embrace both Thai and Japanese cultures relatively holistically. As they appreciate both cultures from the beginning of their migration

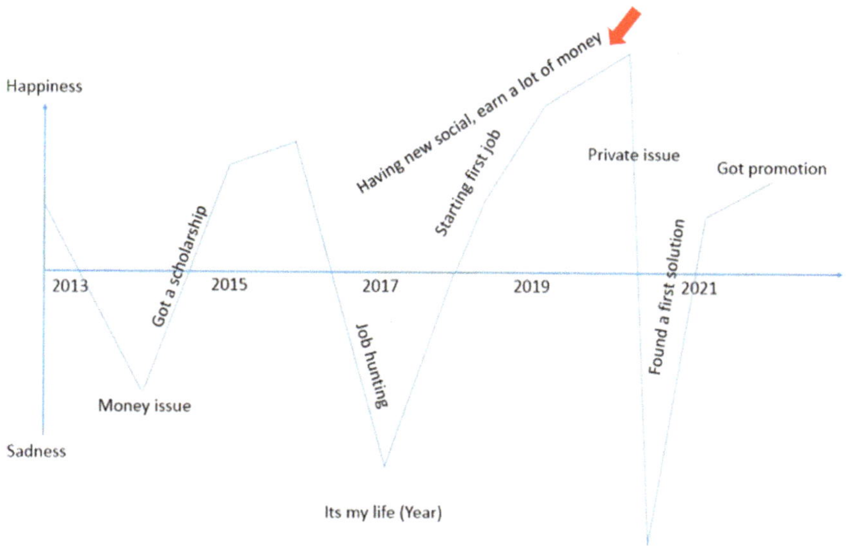

Fig. 6 Ekkapap's life-line

journey, they have never encountered situations where they need to choose or set aside any cultures. In addition, they also express gratitude toward the Japanese culture as it enriches themselves as better *Thai* persons.

> Japanese people always behave well, so when I do something against the social norm, people around me sometimes remind me about dos and don'ts…I think I gradually develop the mindset of selflessness from living in Japan… I feel like it is a positive change that made me become a better person without losing my identity as a Thai (RESP#11).

As illustrated in Fig. 6, Ekkapap's success in socializing with Japanese people and adjustment to Japanese society are catalysts of a rise in his self-esteem. He also pointed out that the personalization of Japanese norms and values has no negative effects on his Thai cultural identity.

5.2 *Narrative on Sense of Self-continuity*

Since life-stories narratives feature many impactful events for an individual's personality development, the way these events are organized in the narratives are crucial to measure the sense of self-continuity. The coherence and reasonability in the narratives clarify the extent to which individuals can recall and relate their past to the present selves, which is a strong indication of self-continuity. Following Camia and Zafar (2021)'s analytical approach, the measurement of self-continuity relies on the global coherence and autobiographic reasoning embedded in the narratives. The measurement of self-continuity relies on the global coherence and autobiographic reasoning embedded in the narratives.

Coherence implies narrators' ability to properly maintain some aspects of an assigned topic within a discourse. Whereas local coherence indicates consistency in meanings between given and repeated utterances, global coherence demonstrates the relationship between repeated narratives and the overall theme of a conversational topic (Strauss Hough and Barrow 2003). Camia and Zafar (2021) further divided global coherence into three kinds; temporal, causal-motivational, and thematic coherence. The temporal coherence indicates a clear explanation and order of life events. The casual-motivational coherence associates the narratives regarding a sense of self-development due to some particular events. Lastly, thematic coherence illustrates the structures and organizations of narratives into themes. In terms of analysis, the excerpts of narratives that fit the descriptions of each coherence are coded under predetermined themes which will reflect the global coherence of overall narratives.

Autobiographical reasoning is interrelated to global coherence. Whereas global coherence deals with arrangement and classification of life events, autobiographical reasoning further elaborates changes and continuity in identity by means of connecting biographical experiences in the life-course. Within autobiographical reasoning, autobiographical arguments are divided into six themes; developmental status, biographical background, lessons learned, generalized insights, formative experience, and turning points (Habermas and Köber 2015; Camia and Zafar 2021). Life narratives were coded for the present autobiographical arguments.

With the assistance of MAXQDA software, the excerpts are assigned under each theme of global coherence and autobiographical reasoning (see Table 2). The content analysis for global coherence and autobiographical reasoning shed light on the degrees of self-continuity and self-discontinuity of each participant.

The content analysis shows that the sense of self-continuity is vivid and consistent with concrete evidence of narrative global coherence and autobiographical reasoning for almost all the participants. However, the sense of self-continuity is questionable for few participants due to unclear global coherence and autobiographic reasoning in their narratives.

> Being in that situation (complaint by supervisors), anybody could lose their mind. I felt unsure about the decisions I made. It was the time that I cried every day nonstop. I didn't even remember exactly what else have I done. It felt so numb and robotic. To be honest, it is even difficult for me to plot this event in the life-line. How to say it is like, I'm sure it was the saddest moment, but the time passed by without much realization, and the details were now quite blurry (RESP#7).

> Then I tried to remind myself when I was here and how hard it was to get the job. Support from my colleagues and family was also very encouraging... I tried not to repeat the mistakes and became more familiar with the job. I cannot say I'm completely fine, but it got a lot easier than before (RESP#13)

The narratives embedded some messages like "I didn't even remember exactly what else I had done" or "It felt so numb and robotic," which signify the identity disruption and confusion. These statements also neither align with any of the three

Table 2 Samples of excerpts under global coherence and autobiographical reasonings

Global coherence	
Temporal coherence	From 2020, here were lots of ups and downs here depending on the situations that I encountered
Causal-motivational coherence	When I got used to the new system, it was not that bad. I think I've kinda got tougher because of the situation
Thematic coherence	I was allowed many day off so I had a chance to **travel** a lot. However, when I needed to get back to **work**, it could be challenging and stressful…The graph went down because I needed to adapt to **new work environment**, but it went up when I **explored new places** with friends
Autobiographical reasoning	
Development status	… I feel like it is a positive change that made me become a better person without losing my identify as a Thai
Biographical background	I cannot forget when a stranger eagerly helped me with paperwork even thought he couldn't speak English. That's why I wished to always help people out
Lesson learned	…Then I realized that calling people by their first name can sound inappropriate, so I've tried to be more careful
Generalized insights	In Thailand, the job opportunities are quite limited. Especially in the field that I'm from, it's quite hard to get a job with a high salary. That's why I decided to leave Thailand and am hopeful for a better life here
Formative experience	I didn't want to stay in Japan for so long because I couldn't speak Japanese and didn't like Japan that much. After I started dating my (Japanese) boyfriend, I changed my mind and I tried hard to stay in Japan so that I could be close to him
Turning points	Yeah, because when I first moved to this city. I was lonely and hardly knew anyone. Until around April 2018, when I joined a

global coherences nor six autobiographical reasonings. For these periods of difficulty, their sense of self-continuity, therefore, can be called into question. Nonetheless, the period that the participants experienced the biographical disruption was relatively short. Plus, they sorted out the way to regain self-esteem and life purpose, which is an important element of autobiographical meaning-making. Their overall narratives, except this part, are rich in global narrative coherence and autobiographical reasoning. Hence, it could be argued that these participants only experienced a temporarily disturbed sense of self-continuity rather than chronic self-discontinuity.

Because of remarkable global coherence, and autobiographic reasoning in the participants' narratives and a slight number of participants experiencing a short period of disturbed sense of self-continuity, this research indicates the participants' success in protecting their sense of self-continuity. The following section discusses how acculturation strategies and a sense of self-continuity are interrelated.

5.3 Correlation Between Acculturation Strategies and Sense of Self-(dis)continuity

5.3.1 Positive Correlation Between Integration and Sense of Self-continuity

The findings reflect an obvious and stable sense of self-continuity among all participants on the one hand and the integrated approach as a prevailing acculturation strategy on the other. In this section, the correlation between the sense of self-continuity and the integrative acculturation strategy is explored. To illustrate, the direct link between the two elements is shown in many excerpts from participants' narratives.

> When I got closer to my colleagues, I was more eager to learn Japanese because I had a clear purpose for learning. I also enjoy sharing my stories about my Thai friends, family, and tradition. I was happy when they looked excited and even wished to go to Thailand some days (RESP#3).

> I don't think I forgot my ethnic background. It's quite the opposite. When Japanese culture has become a part of my life, I often reflect Thai culture in comparison…I miss Thai food when I'm in Japan. I appreciate that Thai people use polite forms to family members, which is uncommon in Japan…I've become a punctual person after many years living in Japan, and I wish Thai people concern more about punctuality (RESP#4).

> While the Japanese are very careful when talking to seniors, Thai people are also taught to always respect elders. Compared to other foreign colleagues, I feel like the Japanese, especially the elders, are slightly happier interacting with Thai people. Perhaps, they share a similar sense of respect for seniority (RESP#10).

All the excerpts above demonstrate the degrees to which participants maintained their cultural integrity. Additionally, they also unveil the participants' endeavors to extend a social network with Japanese people. In this sense, the narratives proved that the maintenance of original culture plays an important role in successful cross-cultural interaction. In a similar vein, the sense of comparison of cultural differences and similarities became a tool for self-reflection and reminder. Therefore, the analysis suggests the reciprocal link between integration as an acculturation strategy and the sense of self-continuity.

Another piece of evidence to prove the correlation between the sense of self-continuity and integration is the global coherence and autobiographic reasoning embedded in the narratives about the integrative approach of acculturation strategy. The area of the life-line below was used to narrate the following story (see Fig. 7).

> In 2019, I moved from Nara to Hiroshima. This was quite a big change in my life because the environment was very good. I made friends with a lot of Thai people. My Japanese colleagues were very supportive. They also taught me nicely about cultures in the organization and in general….I think embracing the Japanese norms makes me a better person somehow… My financial status was also stable. That was the happiest moment (RESP#11).

This participant mentioned social support from both Thai and Japanese people, which helped him nurture psychological and sociocultural well-being. It is also salient

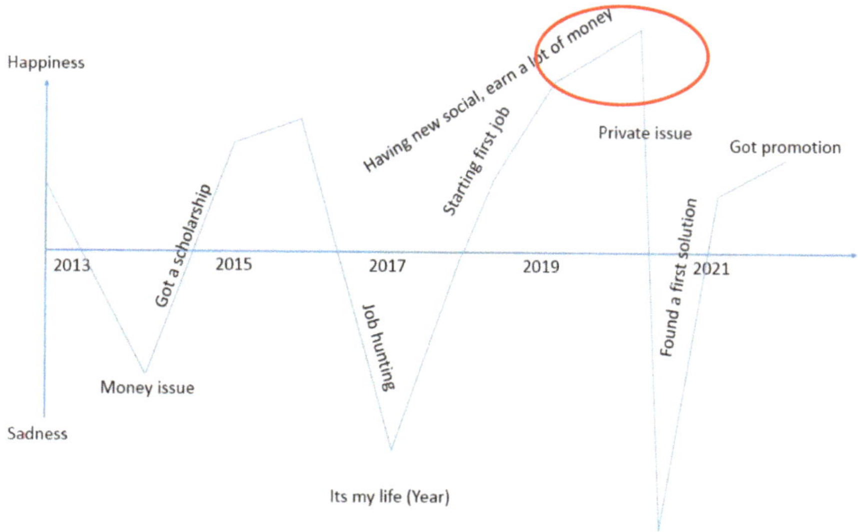

Fig. 7 Sample of a life-line accompanied with an excerpt

that the narrative features thematic, casual-motivational, and temporal coherence, as well as unique features of autobiographical arguments (e.g., turning point, formative experience, developmental status). The analysis points out another aspect of the correlation between integration and the sense of self-continuity.

6 Conclusion

This research was conducted to identify the prevailing and most effective acculturation strategies (assimilation, integration, separation, marginalization) among highly skilled Thai migrant workers in Japan. With the realization of the importance to promote and protect cultural rights among migrants, this study also explored the sense of self-continuity conceptualized through the cultural lens. The correlation between acculturation strategies and the sense of self-continuity was also the other key research objective.

The narrative analysis sheds light on integration as a dominant acculturative approach for twelve highly skilled Thai migrant workers. This integrative approach can be divided into three patterns. Firstly, the change from assimilation to integration similarly arose from a number of participants' narratives. To explain the paradigm shift, assimilation was perceived as a way to gain social acceptance, but it later took a toll on the self-esteem of the participants. Therefore, they utilized the integration approach and were more satisfied with the adaptation outcome. Secondly, some participants were encouraged to strictly follow Japanese norms in the workplace setting, which leads to the prevalence of acculturation strategy. Nevertheless, compliance with Japanese social expectations becomes flexible in the everyday-life domain, which tolerates the continuation of Thai culture. In this sense, the balance between two forms of cultural activation can be referred to as integration strategy. Lastly, unlike the other two approaches, some participants conceive Thai and Japanese cultures as complementary to each other. Hence, they maintain and embrace home and host cultures simultaneously and holistically. As for the self-continuity, the overall robust global coherence and autobiographic reasoning in the narratives of the participants are reflected in this study. Thus, the findings indicate a stable sense of self-continuity for the participants. This study also highlights positive correlation between integrative approach of acculturation strategy and the sense of self-continuity. It is evident that the personalization of both cultures provokes the sense of comparison and reminds the participants of their original culture, which facilitates continuity of their cultural identity.

Usually portrayed as a country with a robust collective ethnonational attitude, discrimination against foreigners in Japan have been a controversial issue that has sparked discussion by a number of scholars (Kim 2011; Meek 2004; Otsuka and Anamizu 2019). However, this study depicts a stark contrast, as Japanese people were referred to by research participants with welcoming and open-minded attitudes toward international colleagues. It is also worth mentioning the extent to which each company differently promotes the discourse of cultural diversity depending on their commitment to traditional Japanese norms.

Some unique characteristics which may explain the positive interpersonal relationship between Thai participants and their Japanese colleagues in this study are high education and Japanese proficiency. Whereas the role of language and education in bridging understanding can be applied to explain this phenomenon, Meek (2004) argued that highly skilled migrant workers could also fall victim to bullying and discrimination in the workplace, regardless of their linguistic skills and educational background. In this regard, education and the ability to communicate in Japanese alone cannot sufficiently explain successful integration for the research participants.

Drawing from the participants' narratives, a more obvious and crucial instigator of successful integration is the concerted endeavor of Thai participants to follow organizational norms and regulations influenced by Japanese culture. Furthermore, as all the participants lived in Japan for a few years as students before employment, they developed familiarity with Japanese culture, which facilitates smooth and effortless adjustment into work society.

Some limitations and challenges have arisen in this research. Despite insightful narratives, the number of 12 participants with similar socio-economic characteristics makes the findings not generalizable. Moreover, this research assumes the attitudes and prevailing norms of a particular group of Japanese people without a direct conversation with them. Research with similar topics has the potential to address these gaps by enlarging and diversifying sample sizes. In terms of intercultural study, the voices that represent all cultural groups can improve the rigor and validity of research. All in all, this study is novel in terms of theories applications which can be beneficial for both academia and policymakers.

This research makes three major contributions. First, the incorporation of self-continuity to understand the acculturation process brings in the cultural emphasis. This theoretical framework can be applied to studies regarding human rights and psychological well-being. Second, whereas self-continuity was dominantly discussed in the sociopsychological discipline, the notion has been reconceptualized in this study to understand cultural identity. In other words, it takes an interdisciplinary approach to extend discussion on self-continuity topics. Last but not least, the findings from this research highlight the necessity of the promotion of cultural rights among migrants not only to safeguard the well-being of the migrants but also to foster a multicultural environment, which can be tremendously beneficial for Japanese aging society. Japanese decision-makers are encouraged to consider the cultural rights protection mechanisms upon the design and implementation of migration policy.

Acknowledgements The author would express heartfelt gratitude toward Prof. Katayanagi Mari for her continued support for this research project as well as twelve participants in this study.

References

Ager A, Strang A (2008) Understanding integration: a conceptual framework. J Refug Stud 21(2):166–191. https://doi.org/10.1093/jrs/fen016

Aichinger T, Fankhauser P, Goodman R (2017) Worsening work conditions and rising levels of job satisfaction? Measuring the happiness of academics in Japan. Res Comp Int Educ 12(2):213–230. https://doi.org/10.1177/1745499917712609

Aoyama K (2009) Thai migrant sex workers: from modernization to globalization. Springer

Arimoto A, Daizen T (2012) Factors determining academics' job satisfaction in Japan from the perspective of role diversification. In: Job satisfaction around the academic world. Springer, Netherlands, Dordrecht, pp 145–165

Becker M, Vignoles VL, Owe E, Easterbrook MJ, Brown R, Smith PB et al (2018) Being oneself through time: Bases of self-continuity across 55 cultures. Self Identity 17(3):276–293. https://doi.org/10.1080/15298868.2017.1330222

Berry JW (2005) Acculturation: living successfully in two cultures. Int J Intercult Relat 29(6):697–712. https://doi.org/10.1016/j.ijintrel.2005.07.013

Bengtsson M (2016) How to plan and perform a qualitative study using content analysis. NursingPlus Open 2:8–14

Cabassa LJ (2003) Measuring acculturation: where we are and where we need to go. Hisp J Behav Sci 25(2):127–146. https://doi.org/10.1177/0739986303025002001

Camia C, Zafar R (2021) Autobiographical meaning-making protects the sense of self-continuity past forced migration. Front Psychol 12:618343. https://doi.org/10.3389/fpsyg.2021.618343

Castles S (2004) Migration, citizenship, and education. In: Diversity and citizenship education: global perspectives, pp 17–48. https://doi.org/10.4337/9781788112376

Chandler MJ, Lalonde CE, Sokol BW, Hallett D, Marcia JE (2003) Personal persistence, identity development, and suicide: a study of native and non-native North American adolescents. Monogr Soc Res Child Dev i–138. https://doi.org/10.7202/018011ar

Crown DF, Rosse JG (1995) Yours, mine, and ours: facilitating group productivity through the integration of individual and group goals. Organ Behav Hum Decis Process 64(2):138–150. https://doi.org/10.1006/obhd.1995.1096

Department of employment of Thailand. *รายงานการวิจัยเพื่อการพัฒนาองค์กรและระบบจัดส่งแรงงานไทยไปต่างประเทศโดยรัฐ* [Research Report for Development of Thai Workers Overseas Deployment Organization and System by Government], Department of Employment of Thailand
Dunkel CS (2005) The relation between self-continuity and measures of identity. Identity 5(1):21–34. https://doi.org/10.1207/s1532706xid0501_2
Economic Commission (n.d.) Economic migrant. Migration and Home Affairs. https://home-affairs.ec.europa.eu/networks/european-migration-network-emn/emn-asylum-and-migration-glossary/glossary/economic-migrant_en#:~:text=Definition(s),material%20improvements%20in%20their%20livelihood
Eliades M, Mansell W, Blanchette I (2013) The effect of emotion on statistical reasoning: findings from a base rates task. J Cogn Psychol 25(3):277–282. https://doi.org/10.1080/20445911.2012.761632
Elo S, Kyngäs H (2008) The qualitative content analysis process. J Adv Nurs 62(1):107–115. https://doi.org/10.1111/j.1365-2648.2007.04569.x
Erikson EH (1968) Identity and identity diffusion. In: Gordon CH, Gergen KJ (eds) The self in social interaction. Vol. I: classic and contemporary perspectives. Wiley, New York, pp 197–206
Ersner-Hershfield H, Wimmer GE, Knutson B (2009) Saving for the future self: neural measures of future self-continuity predict temporal discounting. Soc Cogn Affect Neurosci 4(1):85–92. https://doi.org/10.1093/scan/nsn042
Flannery WP, Reise SP, Yu J (2001) An empirical comparison of acculturation models. Pers Soc Psychol Bull 27(8):1035–1045. https://doi.org/10.1177/0146167201278010
Fontana E (2020) When the main job tasks are perceived to be 'irrelevant in the workplace: the internal uselessness of corporate social responsibility work in Japan. Cult Organ 26(5–6):405–424. https://doi.org/10.1080/14759551.2019.1656632
Fresnoza-Flot A (2021) 'Men are butterflies, women are hindlimbs of an elephant': Thai women's gendered being in transnational spaces. Gend Place Cult 28(5):680–701. https://doi.org/10.1080/0966369x.2020.1752628
Froese FJ (2010) Acculturation experiences in Korea and Japan. Cult Psychol 16(3):333–348. https://doi.org/10.1177/1354067x10371138
Fyvie A, Ager A, Curley G, Korac M (2003) Integration mapping the field volume II: distilling policy lessons from the "mapping the field" exercise. Home Office Online Report 29(03)
Habermas T, Köber C (2015) Autobiographical reasoning in life narratives buffers the effect of biographical disruptions on the sense of self-continuity. Memory 23(5):664–674. https://doi.org/10.1080/09658211.2014.920885
Hinenoya K, Gatbonton E (2000) Ethnocentrism, cultural traits, beliefs, and English proficiency: a Japanese sample. Mod Lang J 84(2):225–240. https://doi.org/10.1111/0026-7902.00064
Hiroshima Prefecture (n.d.) *在留外国人等の状況 (1) 広島県内国籍(出身地)別在留外国 人数の推移*, [Status of Foreign Residents, etc. (1) Changes in the number of foreign residents in Hiroshima Prefecture by nationality (place of origin)]. www.pref.hiroshima.lg.jp/uploaded/attachment/423647.pdf. Accessed 26 May 2023
Hogg MA (2003) Social identity
ILO Brief—International Labour Organization. https://www.ilo.org/wcmsp5/groups/public/---asia/---ro-bangkok/---sro-bangkok/documents/publication/wcms_760374.pdf. Accessed 21 Apr 2023
Jetro (n.d.) Data: attractive local regions in Japan—investing in Japan–Japan external trade organization. ジェトロ. Retrieved 22 May 2023. https://www.jetro.go.jp/en/invest/region/data/hiroshima.html
Ji X, Chui CH, Ni SG, Dong R (2020) Life satisfaction of rural migrant workers in Urban China: the roles of community service participation and identity integration. J Soc Serv Res 46(2):273–282. https://doi.org/10.1080/01488376.2018.1555110

Joiner TE Jr, Rudd MD (1995) Negative attributional style for interpersonal events and the occurrence of severe interpersonal disruptions as predictors of self-reported suicidal ideation. Suicide Life-Threat Behav 25(2):297–304. https://doi.org/10.1111/j.1943-278X.1995.tb00927.x

Kalapong J (2023) Blessing the living spaces of migrants: religious practices of Thai workers in Japan and their meanings. South East Asia Res 1–16. https://doi.org/10.1080/0967828X.2023.2174041

Kamolkhantibowon A (2018) How does the Japanese workplace culture motivate and discourage Thai employees in Japan?

Kim B (2011) Blatant discrimination disappears, but…: the politics of everyday exclusion in contemporary Japan. Asian Perspect 35(2):287–308. https://doi.org/10.1353/apr.2011.0008

Kitiarsa P (2008) Thai migrants in Singapore: state, intimacy and desire. Gend Place Cult 15(6):595–610. https://doi.org/10.1080/09663690802518495

Leong FT, Chou EL (1994) The role of ethnic identity and acculturation in the vocational behavior of Asian Americans: an integrative review. J Vocat Behav 44(2):155–172. https://doi.org/10.1006/jvbe.1994.1011

Leveto JA (2016) Happiness and identity theory. In: New directions in identity theory and research, vol 1

McClelland BP (2020) Identity and agency: majority and minority ethnic voting in new democracies. Columbia University

Meek CB (2004) The dark side of Japanese Management in the 1990s: Karoshi and time in the Japanese Workplace. J Manag Psychol 19(3):312–331. https://doi.org/10.1108/02683940410527775

Milner A, Browitt J (2013) Contemporary cultural theory: an introduction. Routledge

Ministry of Health, labor, and Welfare (n.d.) We are at the door to the super aged society. Ministry of Health, labor, and Welfare. https://www.mhlw.go.jp/english/org/policy/p32-33.html

Ministry of Health, labor, and Welfare (n.d.). *「外国人雇用状況」の届出状況まとめ(令和４年10月末現在)* [Summary statistic of "foreign employment status" (as of the end of October 2020)]. Ministry of Health, labor, and Welfare (n.d.). https://www.mhlw.go.jp/stf/newpage_30367.html. 27 Jan 2023

Morita L (2015) Some manifestations of Japanese exclusionism. SAGE Open 5(3):2158244015600036

Nakajima Y (1997) Politeness strategies in the workplace: which experiences help Japanese businessmen acquire American English native-like strategies? Work Pap Educ Linguist 13(1):49–69

Neuliep JW, Chaudoir M, McCroskey JC (2001) A cross-cultural comparison of ethnocentrism among Japanese and United States college students. Commun Res Rep 18(2):137–146. https://doi.org/10.1080/08824090109384791

Niedenthal PM, Rychlowska M, Wood A, Zhao F (2018) The heterogeneity of long-history migration predicts smiling, laughter, and positive emotion across the globe and within the United States. Plos One 13(8):e0197651. https://doi.org/10.1371/journal.pone.0197651

Ong A, Dominguez VR, Friedman J, Schiller NG, Stolcke V, Wu DY, Ying H (1996) Cultural citizenship as subject-making: Immigrants negotiate racial and cultural boundaries in the United States [and comments and reply]. Curr Anthropol 37(5):737–762

Office of Labour Affairs. The general problem of Thai workers—Office of Labour Affairs royal ... (n.d.). https://japan.mol.go.th/en/situation/the-general-problem-of-thai-workers

Onorato RS, Turner JC (2004) Fluidity in the self-concept: the shift from personal to social identity. Eur J Soc Psychol 34(3):257–278

Osei-Kofi N (2012) Identity, fluidity, and groups: The construction of multiracialism in education discourse. Rev Educ Pedag Cult Stud 34(5):245–257. https://doi.org/10.1002/ejsp.195

Otsuka H, Anamizu S (2019) Japanese university students' difficulty in living and its association with suicidal ideation. Asian J Psychiatr 43:50–52. https://doi.org/10.1016/j.ajp.2019.05.004

Park H (2019) A model of identity development: life-course narratives of North Korean youth resettling in South Korea. J Adolesc 76:162–172. https://doi.org/10.1016/j.adolescence.2019.08.014
Phinney JS (2010) Understanding development in cultural contexts: how do we deal with the complexity? Hum Dev 53(1):33–38. https://doi.org/10.1159/000268138
Piontkowski U, Rohmann A, Florack A (2002) Concordance of acculturation attitudes and perceived threat. Group Process Intergroup Relat 5(3):221–232. https://doi.org/10.1177/1368430202005003003
Raijman R (2013) Foreigners and outsiders: exclusionist attitudes towards labor migrants in Israel. Int Migr 51(1):136–151. https://doi.org/10.1111/j.1468-2435.2011.00719.x
Ruenkaew P (2002) Female Thai migrants in Japan. J Gend Stud 5(22):47–69
Sam DL, Berry JW (2010) Acculturation: when individuals and groups of different cultural backgrounds meet. Perspect Psychol Sci 5(4):472–481
Sedikides C, Hong EK, Wildschut T (2023) Self-continuity. Annu Rev Psychol 74:333–361
Sellek Y (1996) Female foreign migrant workers in Japan: working for the yen. Jpn Forum 8(2):159–175 (Taylor & Francis Group). https://doi.org/10.1080/09555809608721567
Schwartz SJ, Zamboanga BL, Weisskirch RS (2008) Broadening the study of the self: integrating the study of personal identity and cultural identity. Soc Pers Psychol Compass 2(2):635–651. https://doi.org/10.1111/j.1751-9004.2008.00077.x
Schwartz SJ, Unger JB, Zamboanga BL, Szapocznik J (2010) Rethinking the concept of acculturation: implications for theory and research. Am Psychol 65(4):237. https://doi.org/10.1037/a0019330
Smith RJ (1961) The Japanese rural community: norms, sanctions, and ostracism. Am Anthropol 63(3):522–533
Smutkupt S, Kitiarsa P (1999) Khon chaikhop: chiwit lae chumchon khong khonngan thai nai yipun [Surviving in the Marginal World: Life and Community of Thai Migrant Workers in Japan]. Anthropol Globaliz: Thai Exp 1999:13–63
Steele CM, Spencer SJ, Aronson J (2002) Contending with group image: the psychology of stereotype and social identity threat. In: Advances in experimental social psychology, vol 34. Academic Press, pp 379–440. https://doi.org/10.1016/S0065-2601(02)80009-0
Straiton ML, Ansnes TJ, Tschirhart N (2019) Transnational marriages and the health and well-being of Thai migrant women living in Norway. Int J Migr Health Soc Care 15(1):107–119
Strauss Hough M, Barrow I (2003) Descriptive discourse abilities of traumatic brain-injured adults. Aphasiology 17(2):183–191. https://doi.org/10.1080/729255221
Swann WB Jr, Bosson JK (2008) Identity negotiation: a theory of self and social interaction. In: John OP, Robins RW, Pervin LA (eds) Handbook of personality: theory and research. The Guilford Press, pp 448–471
Takamatsu R (2021) Culture and extrinsic moral motives: comparisons of Japan residents and Japanese sojourners in individualistic and loose countries. Cult Brain 9(2):112–127
Tan KS (2018) Traversing the Golden Mile: an ethnographic outline of Singapore's Thai Enclave. J Urban Ethnograph 3
Tian Y (2019) Workers by any other name: comparing co-ethnics and 'interns' as labor migrants to Japan. J Ethn Migr Stud 45(9):1496–1514. https://doi.org/10.1080/1369183X.2018.1466696
Trettevik R, Grindal M (2016) The ideal and ought self-guides and the affective consequences of identity verification. In: New directions in identity theory and research, pp 602–626
UNHCR Global Report 2005—Glossary. UNHCR (n.d.). https://www.unhcr.org/media/unhcr-global-report-2005-glossary
Vink MP, Bauböck R (2013) Citizenship configurations: analysing the multiple purposes of citizenship regimes in Europe. Comparat Euro Politics 11:621–648
Ward C, Kus L (2012) Back to and beyond Berry's basics: the conceptualization, operationalization and classification of acculturation. Int J Intercult Relat 36(4):472–485

Wiese D, Escobar JR, Hsu Y, Kulathinal RJ, Hayes-Conroy A (2018) The fluidity of biosocial identity and the effects of place, space, and time. Soc Sci Med 198:46–52. https://doi.org/10.1016/j.socscimed.2017.12.023

Wolf R, Gan R (2013) Management relations in the work culture in Japan as compared to that of the West. Innov J Bus Manag 2(5):116–122

World Bank (n.d.) Japan's new immigration policy may be a development game changer for South Asia. World Bank Blogs. https://blogs.worldbank.org/peoplemove/japan-s-new-immigration-policy-may-be-development-game-changer-south-asia. Accessed 21 Apr 2023

Zetter R, Griffiths D, Sigona N, Hauser M (2002) Survey on policy and practice related to refugee integration. In: Oxford, European Refugee Fund Community Actions 2001/2002; conducted by School of Planning, Oxford Brookes University

Zhang H (2021) From Thailand to Hong Kong: reinventing womanhood in the journeys of migrant women. South East Asia Res 29(2):231–247. https://doi.org/10.1080/0967828X.2021.1934723

Connecting the Dots: Assessing the Role of 'Women, Peace and Security' Agenda in Conflict Resolution and Peacebuilding in Climate Fragile Zones of the Rohingya Camps in Bangladesh

Mokbul Morshed Ahmad and Robaiya Nusrat

Abstract Peacebuilding is a process often required in the form of activities for both pre- and post-conflicted states. Neo-liberal policy-makers, academicians, and other actors including the UN Peacebuilding Commission often associate peacebuilding with a 'post-conflicted' state. This application of the term is limited to activities taking place during a state of conflict and when reconstruction begins. In some protracted humanitarian crises, peacekeeping agenda as per Women Peace and Security (WPS) norms have not been addressed. Additionally, WPS agenda drops the contextual understanding of addressing gender as intersectional to climatic impasse in hazardous geographies where the refugee camps are situated. Even though WPS have lately recognised climate change as a threat multiplier in international peace and security operation, this adoption in the UN Security Council's Resolutions is still unclear, which is an indication that issues pertaining to 'human development' and 'human security' remains unaddressed. This paper critically assesses literature to identify the nexus between peacebuilding and peacekeeping processes that encompass sustainable development through climate action by addressing gender neutral roles in protracted crises. A case has been assessed from the Rohingya refugee camps in Bangladesh, with a conclusion that considerable interlinkages persist not only within the sustainability framework—economic, environmental, social, and institutional—but also has synergies and trade-offs across economic development peripheries, ecosystem functioning, peace, and conflict management. Blending all functions together in protracted crises, therefore, requires a transition from addressing the complex issue of environmental-economic systems towards positive peace nexus.

M. M. Ahmad · R. Nusrat (✉)
Department of Development and Sustainability, School of Environment Resource and Development, Asian Institute of Technology, Pathum Thani, Thailand
e-mail: nusrat.robaiya@gmail.com

M. M. Ahmad
e-mail: morshed@ait.ac.th

A. Sharifi et al. (eds.), *Bridging Peace and Sustainability Amidst Global Transformations*, World Sustainability Series,
https://doi.org/10.1007/978-981-99-7572-3_13

Keywords Conflict · Women peace and security · Peacekeeping · Peacebuilding · Climate hazards · Eco-system

1 Introduction

For decades, women have been critiqued for their movements in justifying their rights. Mainstreaming gender roles across national boundaries and with international players had been a unique set of challenges for women (Swaine et al. 2019) who had been justifying their rights through various platforms and forums (Banda 2013). Movements have been pivotal in transiting from each conceptual right developed by international institutions (Mack, 2005; Chen et al. 2001); which is, from Woman in Development (WID) to Women and Development (WAD) and finally, Gender and Development (GAD). Have seen conceptual evolutions from one approach to another, generations have aspired for greater equality and empowerment. Sufficient conceptual legacies had been put into play where women's roles have transitioned within development contexts for the purpose of enhancing security and stability Hugh (2003), with such concepts being associated with human rights (Pankhurst 2000).

According to Parlevliet (2011), one thing which is common in each of these historical frameworks is that, these decades long movements have not resolved violations in human rights or acknowledged differing needs for men and women in accessing resources in protracted crises. Conflict drivers may have failed to analyse multiple vulnerability contexts that women face differently (Swaine et al. 2019) than men in transitional economies. This research initially looks into the historical evolution of women's rights from the United Nations (UN) World Conference Report (1980) to national development objectives at a country level, with the addition of the UN General Assembly Resolution (2008) and UNIFEM (2007) reports that raised the issue of human rights-based approaches (HRBA) for the first time. In 2007, UNIFEM introduced HRBA Programming widely into the United Nations Security Council Resolution 1325 for Women Peace and Security (UNSCR 1325 WPS) that was declared in the year 2000. This landmark agenda affirmed women's critical role in addressing security threats, and embarked their equal participation in conflict prevention, conflict resolution, and peacebuilding efforts (WPS 2002).

Alongside, the global community has been experiencing the impacts of climate change at a faster rate than the occurrences of human conflict (Peters et al. 2020). These impacts exacerbated the existing social, economic, and environmental challenges which contributed to various forms of insecurities among and across developed and developing countries (Mawby and Applebaum 2021, Ch. 13). Yoshida (2019) mentions that security concerns linked to climate change induce intersecting and compounding problems of people affected by conflict. Their pre-conditioned vulnerable state further pushes them into various forms of insecurities and deprivations occurring from climate impasse. Some of these have impacts on food-water-energy supply nexuses that increases competition over natural resources, induces loss of livelihoods, creates climate-related disasters, that further results to forced migration and displacements (McCann et al. 2021).

Despite growing recognition of the interlinkages between climate change, peace and security in humanitarian crises, there are only few examples of integrated programmatic approaches that address specific risks at the intersection of climate change and insecurity (Tompkins et al. 2018). Conflict and crisis-affected contexts are more susceptible to climate change, but too often peacebuilding and stabilisation efforts do not consider climate-related impacts or environmental hazards that contribute to mass forced displacements (Stoakes 2019). At the same time, insecurity hinders climate change adaptation efforts (Kuele and Miola 2017), affecting the vulnerable communities and making them even poorer and less resilient. Failure to integrate climate change adaptation initiatives in peacebuilding operations often creates ineffective results (Van Den Hoek et al. 2021).

In conjunction with the above, according to Mawby and Applebaum (2021), world has additionally experienced climate change as a threat which is increasingly affecting human security. Risk drivers from climate change can lead to both direct and indirect insecurities. Miall (2003) addressed the direct impacts on ecosystems triggering secondary risks such as cross-border or internal displacements causing political instability or socio-economic disruptions in host communities. For women and girls in conflicted territories, climatic hazards create intersecting and compounding vulnerabilities (Yoshida 2019; Stoakes 2019), primarily because women and girls experience existing inequalities related to gender roles and potentially struggle from compounding discrimination that theyfaced in their societies. Thus, authors (Marin and Naess 2017; Tompkins et al. 2018; Basu et al. 2020; McCann et al. 2021) concur that these realities make conflict recovery efforts in peacebuilding and protracted crises difficult.

However, atthe WPS agenda which has 10 UN Security Council resolutions working around four pillars: (i) ensuring women's participation in governance and peace and security fields; (ii) preventing conflict and violence; (iii) protecting women's and girls' rights in conflict situations, and (iv) assuring appropriate, just relief and recovery for survivors of conflict (United Nations Report on WPS 2002) reflect the potential of neglecting women's role in addressing climate-related security risks in post-crises (United Nations Report 2000; United Nations Report on WPS 2002). It was only very lately in 2015 that the UN Security Council Resolution 2242 adopted the climate risks and hazards into the WPS agenda alongside other factors taffecting global peace and security (Stoakes 2019; OCHA 2016). Nonetheless, the WPS agenda until now have meagrely recognised climate change as a security risk OCHA (2021) in protracted crises that can be addressed through WPS action plans. In the recent years, despite the fact that the WPS Agenda have considered climate impasse within its action plans, a direct link to climate can only be found in one out of four references.

From the inferences made by authors (Sharifi et al. 2023; Simangan et al. 2023, 2022), it can be concluded that the secondary impacts of climate change on peacebuilding can prominently trigger future conflicts in protracted crises concerning national security and geo-political threats impacting human security; particularly, in areas where peacebuilding efforts have not undertaken climate-induced environmental impacts. It thus complicates the climate security-human security nexus, than it appears, and requires attention across economic and socio-economic contexts in post-conflicted territories to enable positive peace. In order to mitigate environmental degradation in peacekeeping, climate security nexuses with human security response mechanisms are necessary in order to explain causality and associations to security dimensions of peace nexuses which are negatively impacted by environmental and eco-systems degradation in post-crises. Gaps prevail as historical assessments of climate induced environmental degradation and its nexuses with security dimensions are limited in many ways since climate induced risks are unprecedented.

Despite the acknowledgment from international diaspora over growing interlinkages between peace and environmental sustainability (United Nations Report 1992, UNSCR 2019), persistent armed conflicts across the globe have siphoned these initiatives according to Cairns (2000) as cited by Simangan et al. (2021). Authors Bigdon and Korf (2004), Sharifi et al. (2023), Simangan et al. (2023), and Giraldo-Suárez and Rodríguez (2023) have contextualised on environmental stresses exacerbating violence in conflicted-affected or post-conflicted states. Hence, peace and sustainability nexus are complex when stabilizing positive growth through peacebuilding in any protracted crisis as considerable trade-offs are associated with economic, socio-economic and environmental dimensions (Sharifi et al. 2023). The reasons being, in a post-conflict or protracted crisis, the unstable nature of the situation stimulates greater use of natural resources for stimulating economic growth; but over usage leads to conflicting states both in inter and intra borders (Giraldo-Suárez and Rodríguez 2023) imposing conflicts (Van Den Hoek et al. 2021). More recent studies like the IPCC discussed considerable factors influencing peace and environmental sustainability relationships which substantiate the roles of institutions in managing socio-economic contexts affected by climate change induced environmental degradation in order to protect human security. However, the role of WPS in bridging the relationships between environmental sustainability, economic development and eco-system management remains nascent in reality (United Nations Report 2023).

Additionally, climate change in conflicted or post-conflicted territories impose disproportionate risks on women and girls, while the WPS agenda offers a normative framework to ensure women's meaningful participation into climate-conflict or post-conflict interventions (Basu et al. 2020). In this regard, the Climate Change Vulnerability Index [CVI] (2018) has marked Bangladesh as one of the most fragile states suffering from climatic variabilities (droughts, rainfall, landslides, frequent storms, etc.) that exacerbate multiple socio-economic and ecological risks due to environmental hazards in the Southeastern region of the country where the Rohingya refugee camps are situated (National Strategy for Water and Sanitation Hard to Reach Areas of Bangladesh, 2011). The camps are on the hill tracks of Cox's Bazar area where the Bay of Bengal also lies (NAP 2022; Yoshida 2019). The region suffers from loss of natural capital due to high impacts of salinity intrusion from rise in sea level, frequent landslides during monsoons, and susceptibility to erratic rainfalls (Global Climate Risk Index [GCRI] 2021). Nonetheless, the location where the Rohingya refugees reside is known to be one of the largest protracted crises running in the world today;

in a zone that often succumb to landslides due to occurrences of erratic rainfalls or prolonged monsoons (NAP 2022).

Moreover, this area has been marked by Bangladesh's strategic document as one of the hard-to-reach areas given its geographical remoteness, difficulty and inadequacy of communication network to reach the spots according to the (National Strategy for Water and Sanitation Hard to Reach Areas of Bangladesh, 2011). As described by the UN High Commissioner for Human Rights, the Rohingyas of Myanmar have suffered one of the dreadful examples of 'ethnic cleansing' in August 2017 when more than 700,000 Rohingya refugees fled for their lives from Myanmar, crossing the Bangladesh border onto the hilly zones of Cox's Bazaar, which gradually led to the creation of the largest protracted refugee camps; with numbers having escalated today to nearly 100,000 (IASC 2018). Such an involuntary cross-border mass displacement was not experienced by the international diaspora for years from any other conflict-affected territories running in the world today. This, however, prompted the international community to take humanitarian action to protect the refugees, and to be responsive to their particular needs (Ahmad and Nusrat 2022; Van Den Hoek et al. 2021). On the downside, such mass displacements triggered rapid destruction of forests as trees had to be uprooted for infrastructural campsite constructions such as shelter, toilets, water sources, and safe spaces for women and girls who suffered violence (Chowdhury and Arimatsu 2019). As a result, destruction in eco-system was evident from such large-scale displacement (Smith 2020; Islam and Shamsuddoha 2017).

As a result, environmental degradation and resource depletion had enormously impacted the hilly-forest zones of Cox's Bazar (NAP, 2022). Further expansion of new campsites have led to the loss of more than 2000 hectares of forest area. It was also known that constructing campsites disrupted the migration corridors of wild elephants by exacerbating tension between humans and wildlife (Manchanda 2021). As a result of campsites' expansion, biodiversity and eco-system losses impacted the surrounding areas along with the sanctuary, national parks, and conserved forests (Chowdhury and Arimatsu 2019). Even though much has been on the discussion table and global agenda for several years regarding the impacts of mass displacement on biodiversity, the international diaspora has undertaken weak mitigation initiatives through non-binding instruments (Smith 2020). Authors Simangan et al. (2023) and Sharifi et al. (2023) agreed that despite the global occurrence of peace and conflict impact assessments, underlying conflict and fragility drivers in protracted crises have not been dynamically integrated into the peacebuilding response actions, often leading to reoccurrence of conflict that disrupts the sustainability framework. The constant change in climatic impasse and lack of institutional governance in provision of transitional justice disrupts the phase towards sustainable peace (United Nations Report 2023).

In view of the above, the guidelines of the United Nation's High Commissioner for Refugees (UNHCR) mention that all refugee hosting nations should undergo rapid environmental impact assessments, energy feasibility studies and natural resource management plans before allowing large refugee influxes to settle in a host nation (Smith 2020; Carpenter, 2006). However, it is unfortunate that when a sudden outbreak of war occurs overnight like the Rohingya crisis, none of these instruments were put into play (Peters et al. 2020). As an evidence, Bangladesh was least prepared for tackling such a large influx since the government was not only ill-equipped to cope with a sudden and sizeable influx (UNDP Report: Gender and Climate Finance 2017; Marin and Naess 2017); but also the nation itself was submerged in its own development problems of serving the host communities in the region where the refugee settlements are located (Ahmad and Nusrat 2022). McCann et al. (2021), Peters (2020) and Carpenter (2006), shared that host communities and localised territories in conflict-affected areas encountering mass involuntary displacements often have limited knowledge on climate change variabilities.

Thus, there is little understanding on how climate related disasters interact with gender, women and girls in hazardous geographies (Van Den Hoek et al. 2021). It is also less known how communities build resilience under these challenging circumstances (Chowdhury and Arimatsu 2019). Even with the numerous research done pertaining to the WPS's efficacy and effectiveness (Basu et al. 2020), few relationships have been attributed on the nexus of gender-climate-conflict affected areas and the WPS framework's usability within this nexus of affairs for containing peace. Furthermore, there's also little interest in assessing if the framework is potentially ready to take those challenges. Based on the current literature, several authors have shown WPS's significance in the middle-eastern and Afghan conflict-impacted zones where rights of women who had been forcibly displaced were reinstated. According to Peters et al. (2020), Rehn and Johnson Sirleaf (2002), Paffenholz (2004), Pankhurst (2008), Puechguirbal (2010), the role of WPS seemed off-tracked when it came to countering situation like the Rohingya crisis. Gender-based violence (GBV) during and after conflict is still a common phenomenon despite 'years' of WPS resolutions that recognised this problem through effective mechanisms for protecting human security from violence (Diaz 2010; Agarwal 2015; Banda 2013; Austin 2014).

Furthermore, few other literatures revealed that WPS has considered local knowledge from women and girl refugees to design safe spaces inside the camps for countering varied forms of sexual and gender-based violences (GBV) (Smith 2020; UNDP Report: Gender and Climate Finance 2017; McCann et al. 2021). In OCHA (2016), the resolutions incorporated particular needs of women living in refugee camps; for example, improved and safe sanitation facilities, adequate camp illumination, equal food distribution systems and healthcare, education and socio-economic and livelihood opportunities (Islam and Shamsuddoha 2017). However, despite such rule of law in place, certain aspects of this framework are missing in terms of implementation in the Rohingya refugee camps, where refugees experienced atrocities across most spaces and under the current structure within the campsites, and with climate-environmental degradation being a nexus problem in sufficiently addressing gender roles in protracted crisis (UNHCR Report 2021). Accordingly, the author came up with the research objective which is to assess the role of United Nations Security Council Resolution [UNSCR 1325] WPS discourse or action plans at the national level in the failure to address the gender dimensions of climate-environment conflict, and how this framework could have been implemented in synergy with the peace-sustainability nexuses in protracted humanitarian crises like the Rohingya refugee camps.

2 Methodology

The author cited literature and empirical studies which are based on journals, books, intergovernmental reports, and national data pools in order to provide a thorough assessment of the theoretical components that supported debates on the WPS agenda. The literature and empirical studies were assessed to understand the role of gender within climate-environmental degradation nexus in peacebuilding and post-conflict situations where management of economy and ecology is critical. Assessments were evidence-based using a set of guidelines for reporting through systematic reviews and meta-analyses as found in international laws and resolutions. A further discussion was based on case scenarios from the field that have argued against what the meta-analyses in the literature reviews have found with regard to the UNSCR 1325 WPS discourse. The result of implementing the WPS framework in any conflict or post-conflict state was the promotion of sustainable peace. But from synthesis of various studies, persistent gaps were found when addressing the climate-induced environmental degradation through enabling peace measures that encapsulate gender dimensions in peacebuilding processes.

3 Literature Review

3.1 Historical Evolution and the Need for Women Peace and Security (WPS) Framework

The early 1980s saw a considerable rise in landmark theory—Women in Development (WID). According to the United Nations (UN) World Conference Report (1980), it considered national development models from perspectives based on women's views alone. However, the rationale focused more on female economic activities, without shedding light on women's existing role in economic affairs (Miall 2003). Critics recognised the need for acknowledgement of women as agents and contributors to economic development, different from men, while experiencing social change (Nussbaum 2000). Years later, further research considered women's extended experiences in economic and social agenda that evolved into the concepts of *Women* and *Development* (Chen et al. 2001). However, this theory too was less successful in enabling a transformative effect, and ignored gender neutrality that continued to have negative structural effects on women (Diaz 2010).

As Miall (2003) noted, later years saw increased importance of the Gender and Development concept. It saw women as agents of change in all pillars of socio-economic peripheries rather than passive recipients of induced state/cultural laws (Hugh 2003). The concept re-examined and recalibrated gendered roles with social structures to connect more with institutions (Brown 2009). The author also mentioned that it helped women's groups not to be dominated by the state, elites, religious leaders, and delegated 'rights' which otherwise should have been a state law for all. It was being less challenging in declining structural discrimination where dynamics of power transfer were visible. This change pertained to the jurisprudence of feminist human rights (Nussbaum 2000).

Concurrently, in the millennial era, the UN WPS (2002) came to being and brought women's 'right to development' in conflict and peacebuilding stages. This progress was achieved via the notion of 'gender and development' as a human rights-based approach within international development where women were recognised alongside men as equal participants (UN Report, 2002). The UN General Assembly adopted the Declaration of the Right to Development (UN WPS 2002), but criticisms prevailed as it broadly expressed or ensured the active role of women in dispossession, dislocation, and conflict. The UN General Assembly's declaration statement presented a list of human rights violations that states were required to address to facilitate development. Interestingly, this did not include sexism or sexual discrimination. Hence, it appeared that women had minimal engagement at both national and international levels, lacking participation in dialogues or having a voice in decision making (Brown 2009).

In view of the above, economic development theorists in the post-millennial era have prioritised Sen's Capabilities approach[1] to human development which is based on his book Development as Freedom. According to Sen (1998), development models focused on *human* participation, security, their socio-economic sustainability, and human rights towards dignity. On the other hand, the human rights-based approach (HRBA) gained sufficient momentum during the terminal years of MDGs (Millennium Development Goals) in 2015, which was more of a conceptual framework for securing human development based on international human rights standards (Goonesekere 1998; Nussbaum 2000; Agarwal 2015). Furthermore, Nussbaum (2000) aimed to analyse inequalities underlying development constraints, as well as redress discriminatory practices and unjust distributions of power that impede development in conflict-affected regions.

Yet evidence shown in Andreassen and Marks (2006) assert that neither the human rights-based approach nor the right to development framework have consistently analysed rights in a gender-sensitive way in transitional protracted states. Andreassen and Marks (2006) represented their views in a food crisis situation in North Africa where practitioners have acknowledged the specific impact of the crisis on poor families, but failed to assess the role of women in food sourcing, food preparation, or the self-sacrifice of choosing not to eat so that children and other family members could. This seemed to be a startling omission of WPS Framework, given the centrality

[1] (a) to lead a long and healthy life; (b) to be knowledgeable; and (c) to have access to the resources needed for a decent standard of living.

of women's role in food production and availability in both post-conflicted and non-conflicted zones. It also indicates that the sustainable peace notion can be impeded, leading to conflicts when women's key contribution is ignored in potential economic sectors, thus jeopardizing human security (Giraldo-Suárez and Rodríguez 2023).

3.2 *Reclaiming Women Peace and Security (WPS) Agenda in Climate-Conflict-Environmental Degradation Nexus*

The OCHA (2016) report revealed that people globally are more displaced by weather and climatic variabilities than by conflict. It estimates 282 natural disasters annually across the globe creating on average 24.2 million new displacements, either cross border or internal (UNDP Report: Gender and Climate Finance 2017). United Nations University (UNU) 2021's report on Development Framework in conflict-affected climate fragile states mentions that low and middle-income countries are five times more susceptible to frequent weather changes and natural disasters than people in the developed world. While global platforms towards international peace and security is centred around conflict, the constant dynamic geopolitical pattern in the world today interlinks climate-environment and conflict, raising concerns on human security and disproportionately driving deprivations (Stoakes 2019; Tompkins et al. 2018; Parlevliet 2011).

Given the interlinkages and compounding vulnerabilities in fragile, conflict-affected climate vulnerable territories, the WPS agenda provides a critical entry point through which new answers to this complex crisis could be explored and crafted (Kuele and Miola 2017). In 2016, the UN Environment Programme (UNEP) adopted a resolution (better known as Resolution 2/5) to protect the ecosystem in armed conflict that was signed and acknowledged by all member states at the UN Environment Assembly (UN Security Council Resolution [UNSCR] 2019, 2020). In particular, the resolution recognises the "critical importance of protecting the environment at all times, especially during armed conflict, and of its restoration in the post-conflict period, including from the unintended collateral impacts of human displacement resulting from armed conflict" (UNSCR 2019, OCHA 2016).

Since the global community started putting up in the radar the climate-conflict parameters (McCann et al. 2021), protecting eco-systems turned out to be an integral part of humanitarian organisations working in climate-conflict displacement spaces (OCHA 2021). As mentioned earlier, 'Relief' which is one of the pillars within the WPS framework recognises the protection of environment from degrading due to mass displacements (Smith 2020; Puechguirbal 2010; Marin and Naess 2017; Rehn and Johnson Sirleaf 2002). It also counters the fact that host territories will face some degrees of economic-environmental degradation while accommodating new migrants (UNHCR Report 2021). Peters et al. (2020) argued that, if protected, refugee rights can be violated when resources are restricted from being used for optimal living and livelihoods. Authors Chirambo (2021), Manchanda (2021), Islam and Shamsuddoha (2017) mentioned that refugees are disproportionately impacted in their everyday lives resulting to socio-economic conflicts with host communities (Ahmad and Nusrat 2022).

From an example of a Somalian refugee camp, climate variabilities such as shifting precipitation patterns led to water scarcity and contributed to human insecurities that triggered higher rates of interrelated violences (Chirambo 2021). The author further stated that the situation harmed women more as they were involved in the process of sourcing and conserving water. In these arid places where livelihoods are crunched due to water insecurities, girls are married off at an early age for families to cope with livelihood insecurities (Chowdhury and Arimatsu 2019; Giraldo-Suárez and Rodríguez 2023). As a result of climate stressors within conflict-affected displaced communities, sexual violences are higher (McKinnon et al. 2016; Ní Aoláin et al. 2011). For some of these reasons, in 2019, the WPS Index ranked Sierra Leone at 153 out of 167 countries, with 28% of women having experienced intimate partner violence and over 50% of women felt unsafe in their communities. Even though the WPS framework was implemented in some of these fragile geographic spaces, sustainable peace have not been necessarily synergised with the 'Social' pillar within the Sustainability Framework.

The situation is similar with impacts from climate change that mirror similar vulnerabilities that women face in conflict situations (Chowdhury and Arimatsu 2019). Accordingly, in recent years, the WPS Framework started developing National Adaptation Plans (NAPs) which are important mechanisms for supporting refugee voices, especially in women's participation in addressing loss in eco-system (Smith 2020; Chowdhury and Arimatsu 2019) as a result of involuntary displacements mostly in the marked climate vulnerable zones (OCHA 2016). In order to achieve sustainable peace and human security, NAPs consider environmental degradation and loss in eco-systems of specific geographies which are exacerbated by climate variabilities (Marin and Naess 2017) an issue for consideration withrelevance to multiple agenda of the WPS system (OCHA 2021).

3.3 The Impacts of Climate Change are Not Gender Neutral in Protracted Crises

The research handbook on New Directions from Women Peace and Security by Basu et al. (2021) shows that global climate change impacts in post-conflicted territories are creating an uneven burden on women refugees. Just as conflict-affected zones are not gender neutral, climate fragile zones in conflict impacted states create double disparity for women (McCann et al. 2021). The reason being their pre-existing vulnerabilities and patterns of discrimination rooted in structural inequalities adds up with the lack of adaptive capacity for bouncing back from frequent stressors (Smith 2020; Peters et al. 2020; UNDP Report: Gender and Climate Finance 2017). Thus, refugees are forced to be displaced, compounding vulnerability and greater inequality in protracted crises (Chirambo 2021). The Georgetown Institute for Women, Peace and Security and Peace Research Institute's WPS Index (2021) justified that gender gaps in securing employment and wage parity amongst displaced women are greater. WPS Index (2021) further estimated that globally displaced men had 90% more opportunity towards employability than women; and disparities in refugee women were 150% higher in Nigeria, where about only 15% of displaced women could fit themselves into the job market.

It is not new to the world that employability markets are highly segregated by gender divide (Bunch 2004; Pankhurst 2008; Puechguirbal 2010; Parlevliet 2011; Banda 2013). In protracted crises, the problem situation is burdensome as refugee women juggle with meagre experiences, low literacy rates, language barriers, mobility constraints, unpaid care responsibilities, etc. (Basu et al. 2021; Islam and Shamsuddoha 2017). All of these intersecting situations compound the susceptibility of refugee women to being left out from economic opportunities, posing threats toward conflict situation not only within refugee camps, but also with the local inhabitants (WFP Impact Assessment Report 2022). Comparisons between displaced women and host community men exposed even starker gaps in position of women in economic affairs, highlighting the multiple effects of displacement and gender inequality, as in the Rohingya refugee camps (Ahmad and Nusrat 2022). From this inference, it can be concluded that the 'Economic' pillar within the Sustainability Framework have considerable synergies within the WPS Framework for ensuring peace and *economic justice* in protracted crises (Giraldo-Suárez and Rodríguez 2023).

Similar examples were cited by other authors like, in Ethiopia, the share of employed host community men was almost three times the share of employed refugee women (Chirambo 2021). Findings from the WPS Index (2021) suggest that even in countries where displaced women are legally permitted to work, it is an institutional norm for them to face discrimination in wages, position, and option in addition to regulatory barriers. Such impediments affect the economy at large (McKinnon et al. 2016). If gender gaps in employment were removed, most of refugee-hosting nations could have potentially contributed to national GDP. An estimation by Stockholm International Peace Research Institute (SIPRI) Multilateral Peace Operations Database (Smith 2020) shows that USD 1.4 trillion annually is lost from global GDP when refugees (especially women) are off-tracked from contributing in economic opportunities.

Authors Ní Aoláin et al. (2011), Lind (2011), and Agarwal (2015) have well documented that globally, women are disproportionately treated as shown in the percentage of unpaid labour, including provision of household services, such as in ensuring water and fuel for their families. In conflict-affected protracted crises, women spend substantial time in managing household chores and countering climatic pressures which increase their exposure to GBV (George and Shepherd 2016). Facing domestic violence and GBV has implications on women's progression, economic empowerment, and their ability to pursue paid work (Banda 2013; Brown 2009). In addition, girls have lesser scope of school attendance, and a primary factor of dropping out of school is the possibility of facing harassments on their way to commute.

According to Lind (2011), women in Guatemala dedicated at least eight hours per day searching for water to sustain their families. They spent almost double the time and distance required than normal when the place was not impacted by climate-conflict fragilities.

Despite a growing evidence base supporting the fact that inclusive approaches generate better outcomes, the WPS Index (2021) reports that only a meagre 0.01% of Official Development Assistance (ODA) is allocated for gender-responsive climate interventions in protracted humanitarian crises (OECD 2020). The role of WPS in localised humanitarian bases and operations can play a central role in these efforts to build a more secure gender-responsive climate funding (OCHA 2021) for sustainable peace.

4 Findings and Analysis

4.1 *Missing Links: WPS, Climate and the Rohingyas in Protracted Crisis*

The adoption of Resolution 1325 on women, peace, and security, better known asWPS-1325 is still considered a landmark achievement in relation to the roles women played not only as activists, but also as peacebuilders in protracted crises. Despite of more than 20 years of its implementation, considerable weakness within the framework remained, hindering successful execution and localisation of the core pillars (prevention, protection, participation, and peacebuilding and recovery). When a global tool is used locally, it lacks integration with the political-military and socioeconomic structures primarily due to lack of capacity of the people responsible of delivery. Secondly, such global frameworks are often handed down from international peace and security agenda which the locals grapple to enact upon (George and Shepherd 2016). When it comes to consideration of climate change and ways to implement in humanitarian/protracted crises, the WPS agenda can still be seen with scepticism. The framework requires a thorough assessment if it can be aptly executed proactively and pragmatically as a long-term sustainable solution that is inclusive of global peace and security nexus (Basu et al. 2020; Peters et al. 2020).

Even though the United Nations recognised climate change as an important threat multiplier that risks human security in fragile contexts (climate variable and hazardous), little has been seen from the WPS agenda with regards to the impacts of climate change in geographies faced with extreme weather events (OCHA 2016; OECD 2020). For example, the role of WPS was invisible in the most recent Syrian earthquake that occurred in February 2023 despite the huge amount of humanitarian aid from international non-government and inter-governmental organisations flooding in to support the victims. It leads to thinking whether the potentiality of this normative framework has enough synergy with other global frameworks in enabling peace and sustainability. Additionally, few literatures have proven that it is highly risky in places which are experiencing rapid urbanisation. Displacement due to conflict sometimes creates more floating population in the urban areas of host nations. This does not only damage people's economic opportunity, as refugees and locals start competing against each other; but also damages the property of the local inhabitants. This also leads to creation of potential disruption of natural resources and ecosystems which exacerbates strains on local reserves, driving future conflicts and lack of human security.

With the above as the backdrop, the Rohingya refugee camps were seen lacking the intervention of the WPS framework despite receiving much global attention for tackling a sudden onset of humanitarian crisis (IASC 2018). The literature review above highlighted the lack of attention to human rights impeding human security. Human security, for the purpose of this research, has not been limited to conflict, but also to multiple socio-economic and climatic problems arising and risking human lives in protracted crises. The lack of WPS presence in Rohingya protracted crisis had been driving insecurities that impacted people regardless of gender. Within the Rohingya camps, men are unable to play their usual roles as wage earners; and there is a shift in gender roles or a shift in power dynamics, where women are the sole bread-earners. This is a primary cause of women experiencing domestic violence (UNHCR Report 2021; IASC 2018). It comes across as an inherent nature of male heads of households, where women are not expected to earn for their families and be financially independent.

Men without sufficient economic opportunities are often outweighed and frustrated by the rapid manifestation of crisis scenarios, resulting to criminal activisms within the camps (Ahmad and Nusrat 2022). Within the Rohingya camps, women are generally out of their homes and working on camp development initiatives led by humanitarian organisations (UNHCR Report 2021). The male heads of single-headed households are challenged with unique needs as they are often void of skills in maintaining household chores, cooking skills, or in serving as caregivers for their young children, which are often seen as stereotypical responsibilities exclusively assigned to women and girls (WFP: Impact Assessment Report 2022). Yousuf (2022) also agreed to the argument made above on the multiple vulnerability contexts, that in traditional conservative societies, male patriarchy justifies women's exclusion from decision making in family affairs or in public spheres. Similar to the situation in Rohingya camps, women's participation in peacebuilding processes is seldom seen.

Hence, the discussion evolves, highlighting that women and men should be equally included during humanitarian actions which benefits the entire community during any protracted crisis. For instance, Ahmad and Nusrat (2022), Rehn and Johnson Sirleaf (2002), and OCHA (2021) portrayed that both young men and women are among the first responders to a crisis, and playing significant roles that contribute to the resilience of their families and communities.

4.2 Localising WPS by Improving Women's Role in Governance

Women are key actors in addressing drivers of insecurity and fostering peace and stability in conflict-affected territories. Analysis from WPS Index Report (2021) estimated that across 170 countries, those with better scores were more peaceful and stable. The report also covered few countries that were conflict-affected and experiencing protracted crises. Given that climate change is a security threat to human lives, the report also measured those countries which score better but are less vulnerable to climate change and are better prepared to respond to climate variability. These further underscores why looking at these issues together is critical to effective action and how WPS, being a catalyst for women, can accelerate the progress on climate-security, peace-security, human-security and sustain gender equality. Thus, the tension prevailing in the Rohingya refugee camps is due to the competition over economic opportunities or the use of natural resources; and indulging in host-refugee tensions could have been curtailed through the enactment of WPS agenda in the local governance structures (UNHCR Report 2021).

As mentioned above, the geographical location where the Rohingya camps are situated is one of the most climate vulnerable zones, often triggered by landslides or erratic rainfalls. Degradation of natural resources is constant as deforestation and quarrying off hills for creation of spaces for the refugees all add up to the climate change induced ecological imbalances, risking lives and deterring sustainable peace. To improve various determinants of conflict in a protracted zone, WPS's visibility in the area of optimal management of economic resources could have created a significant impact on peace. It is when resources are optimally managed that locals and refugees do not compete and ecology is balanced. Additionally, when men and women refugees do not compete on wage bargains, just peace is enabled.. Women's participation for sustaining peace security efforts could have been successfully addressed when climate-related security risks where acknowledged through implementation of WPS framework. If WPS was applied in Rohingya crises-protracted areas, a locally contextualised framework could have been developed through the WPS national action plans with a focus on women's governance on climate security. In light of this, evidence suggests that inclusion of women at all levels of governance while designing the national action plans is critical for setting up long-term development models. This is especially true when facing climate change, since women play a primary role in natural resource utilization and management and are often first responders to natural and human-induced disasters. However, the gender directeded impacts of climate security risks were absent from climate-environmental degradation debates.

The research study of Georgetown Institute for Women, Peace and Security (2021) highlighted the women in local governance who contributed to building resilience to climate-fragility risks in conflict-affected territories. Their work that has shown results, including women at various levels of governance in fragile contexts (countries most vulnerable to climate change impacts) led to protection of environment and maintained peace within refugee communal spaces. Even though widespread structural inequalities persist, some of these positive changes were seen in post-conflicted state of Nepal and Western African nations who successfully bounced back from long years of civil wars, and where the success of WPS can be acclaimed. The inclusion of women at all levels of decision-making is key to achieving the Sustainable Development Goals, especially in agenda-2, 5, 13, and 16.

5 Discussion: Climate Impacted Conflict Zones are Primary Drivers of Insecurity that Hinders Sustainable Peacebuilding—Multiple Conflict Drivers in the Rohingya Refugee Camps

Since 2002, the UN Security Council had been widely acknowledging the fact that, displacements are not only augmenting refugee vulnerabilities, but also exacerbating pre-existing inequalities in places where refugee camps are situated. In recent years, conflicts have been quite rampant, requiring new gendered risks to be balanced off with climate-oriented specific responses (George and Shepherd 2016). In the situation of the largest protracted humanitarian crisis running in the world today, implementation of WPS resolution and international policies mentioned above had been bleak and widely unaddressed. Even worse, intergovernmental institutions were unable to play a significant role in repatriation, where the lives of these refugees seemed to be in a limbo—being stuck for years in the camps without access to economic, social, or political opportunities (UNHCR Report 2021). A protracted territorial dispute that has evolved from a conflict state in Myanmar had eventually developed a risk of becoming a conflicted *refugee* zone in Cox's Bazar, Bangladesh, given the current trends of violence seen across any gender and children (Islam and Shamsuddoha 2017; Chowdhury and Arimatsu 2019).

However, humanitarian responses provided a strong foundation for extensive diplomatic and political efforts across international partners, including in the UN General Assembly and Security Council, maintaining international attention on the plight of the Rohingya Response (UNSCR 2020). Yet, the pursuit of accountability and finding solutions in Myanmar, and solidarity of international community together with Bangladesh in helping manage this crisis still succumbs to failure at the negotiation tables. The case of Rohingya refugees depicts a reality of protracted crisis remaining unstable even with the practice of theories and the rule of law for containment of peace and security. At the current space of time when this paper is drafted, the protracted crisis still debates between 'repatriation' and 'camp relocation'. Furthermore, while WPS was built for sustaining peace, human dignity across gender norms lies in limbo with such debates pertaining to the Rohingya space. There has been arguments by international development partners against camp relocation to a zone which is more climate vulnerable and where lives of these refugees can be more at risk. However, the area where the camps are now situated happens to be one of the most vulnerable territories in the region from the time settlers moved in 2017. Clearly, the trade-offs to human security are minimal in either of the two places.

From the studies of Parlevliet (2011) and Rehn and Johnson Sirleaf (2002), it was understood that the term '*security*' traditionally meant protection of people's sovereignty within the territory and proclamation of military might. States often failed to heed the '*structural violence*' of poverty, inequality, injustice, and human rights abuse that are often underlying causes of violent conflict and insecurities people face while coping in protracted crises zones. Swaine et al. (2019) further argued that while there has always been a legacy of not acknowledging refugee rights, women as a subset of the wider group had less access to political and economic participation. Even more, they are considered illegitimate and incapacitated for voicing their rights, which is similar to what is happening within the Rohingya camps. When it comes to accessing natural resources within the camp boundaries, they have the primary role of securing fuel and energy sources for their families, in which women often encounter violence while travelling long distances. It can be inferred that sustainable peace can offset all other four pillars of sustainability framework as not interacting with each other proficiently.

According to Coomaraswamy et al. (2015), the international community adopted a comprehensive normative framework to counter sexual violence in conflict. Additionally, the protection mechanisms that come into operation are designed for hugely complex scenarios that take into account the gendered dimensions of displacement, but yet unable to counter climatic risks that drive threats to socio-economic and human security contexts similar to the case of the Rohingyas. To treat something as a 'security' issue has historically meant to legitimise the usual political process and state borders in order to manage security. However, this paper expands the horizon of security agenda, onto the gendered roles of climate change impacts and how gender norms address climate-induced ecological imbalances towards maintaining human security and positive peace (Peters et al. 2020; Sharifi et al. 2023).

WPS is uniquely framed to capture climate change-related secondary impacts to socio-economy such as migration. The framework bridges and integrates two key dynamics: security and gender. Furthermore, WPS is within the mandate of the United Nations Security Council, a governance body with significant normative and policy power. International migration frameworks, national immigration policy, and humanitarian resources are largely unprepared for climate change-related migration flows, creating significant threats to future humanitarian challenges, national security, and border management processes. WPS's entrenchment in the Security Council provides an opportunity to integrate both climate and gender-oriented responses by overcoming challenges to human security as discussed above.

The WPS agenda sits within the United Nations Security Council that creates obstacles for localised implementation. The agenda has been critiqued for being too state-centric, lacking accountability for implementation, and being heteronormative. The framework does not exist in a vacuum; it is not the only framework needed to address climate change or climate change-induced displacements. However, it is a key framework situated at the heart of security policy within the accountability of international community.

The WPS Agenda was slow to recognise climate change as a security risk for women and girls, until the 8th resolution in 2015. It was recognised as relevant to climate change among other secondary factors including a rise in violent extremism and the global nature of health pandemics. Meagerly though, the global context of peace and security, and the subsequent need to recognise WPS was only a "cross-cutting subject". Additionally, as mentioned above, the NAPs under the WPS Agenda have been dysfunctional, not localised as per conflict-affected context, and only 17 countries out of 80 had climate change induced environmental degradation integrated into their nationally adopted WPS-NAPs. In the context of the Rohingya refugee camps, scope has remained relatively unexplored and relatively missing.

6 Conclusion and Recommendations

Environmental degradation is inevitable in the areas of forced (mass) displacements and where conflict strikes. It was mentioned earlier that all 193 states of the UN member countries have recognised protecting their eco-systems as a high priority focus in conflict-affected territories. This gives a good impression that countries are engaging in global partnership due to their commitments to the UN Sustainable Development Goals (SDGs) #16. While global partnerships and addressing SDG #16 (Peace, Justice, and Strong Institutions) are well adhered by all communities, it is pertinent to ensure that a gender perspective is fully integrated into the international rule of law within the spectrum of protecting environmental policies in geographies with protracted crises camps. The degradation of the ecosystem has severe short- and long-term adverse consequences that affect women and men differently with respect to health, livelihoods, security, dignity and wellbeing.

This is recognised in the UNEP Resolution 2016 which draws attention to the "specific negative effects of environmental degradation on women and the need to apply a gender perspective with respect to the environment and armed conflict". Degradation of environment and climate variable impacts are inter-related in the Rohingya spaces. The higher the losses on ecosystems and degradation of the forest areas, the greater will be its exposure to frequent weather and climatic changes. Structuring of WPS-NAPs should continue to improve how they address climate change and climate-related security risks. It is important to consider how climate change is framed and acted on. Specifically, climate-related security challenges for women and girls are framed in a way that promotes their participation in preventing and addressing security risks, instead of focusing only on protection. Climate-related security risks should therefore be addressed comprehensively. Policy actions should be aimed at addressing climate change impacts as a threat-multiplier to women's vulnerability. Ecological imbalances from overuse of economic resources, and further indulging in conflict by competing over common resources would likely not resolve the circumstances. Instead, women's meaningful participation can help overcome challenges by addressing climate-related secondary risks to security across the pillars of the WPS framework. Leveraging existing frameworks in global peace and security efforts and accounting for the unique impacts of conflict and crises on women by integrating climate security into WPS is a critical step to successfully address the security threats posed by climate change. All the more, these efforts to address the climate–environmental degradation crisis and securing sustainable peace towards human security will not reach full potential without scaled up funding. It is important to ensure that mechanisms are in place so resources are utilised for protracted communities most affected by climatic impacts.

References

Agarwal (2015) Conflict zones. In: Walby S et al (eds) Stopping rape: towards a comprehensive policy. Policy Press, London, pp 173–189

Ahmad M, Nusrat R (2022) Human trafficking and crimes in the Rohingya refugee camps in Bangladesh. In: Present and future paradigms in human trafficking. Chapter. 3. IGI Global

Andreassen BA, Marks SP (2006) Development as a human right. Legal, political, and economic dimensions. Harvard Series on Health and Human Rights

Austin B, Fischer M, Giessmann JH (eds) (2014) Advancing conflict transformation: the Berghofhandbook II, pp 377–404

Banda F (2013) Women, human rights and development. Chapter 11

Basu S, Kirby P, Shepherd LJ (2020) Women, peace and security: a critical cartography. In: Basu S et al (eds) New directions in women peace and security. Bristol University Press, UK, Bristol, pp 1–28

Bigdon C, Korf B (2004) The role of development aid in conflict transformation: facilitating empowerment processes and community building aid in conflict. Berghof Res Center Const Conflict Manag Jstore

Brown MA (2009) Security, development and the nation-building agenda—East Timor. Confl Secur Dev 9(2):141–164

Bunch C (2004) A feminist human rights lens. Peace Rev 16(1):29–34

Carpenter RC (2006) Recognizing gender-based violence against civilian men and boys in conflict situations. SAGE Publ 37(1):83–103. https://doi.org/10.1177/0967010606064139

Chen GB et al (2001) Activating human rights and peace: theories, practices and contexts. In: Chapter 14. Women, Peace and securing human rights

Chirambo D (2021) Can social innovation address Africa's twin development challenges of climate change vulnerability and forced migrations? J Entrep Innov Emerg Econ 7(1):60–77

Chowdhury M, Arimatsu L (2019) Reclaiming the WPS agenda: it's time to talk about the elephant in the room. In: Gender research and peace. LSE Research Centre, London, LSE. https://blogs.lse.ac.uk/wps/2019/09/27/reclaiming-the-wps-agenda-its-time-to-talk-about-the-elephant-in-the-room/

Coomaraswamy R (2015) Preventing conflict, transforming justice, securing the peace: a global study on the implementation of United Nations security council resolution 1325. UN Women, New York

Climate Vulnerability Index (2018) Local government initiative on climate change project: report. UNDP, Bangladesh, Dhaka

Diaz PC (2010) Women's participation in peace negotiations: connections between presence and influence. 1325 + 10: women count for peace. UNIFEM

George N, Shepherd LJ (2016) Women, peace and security: exploring the implementation and integration of UNSCR 1325. Int Polit Sci Rev 37(3):297–306

Giraldo-Suarez AI, Rodríguez NL (2023) Intergroup selection as a way to peace and sustainability. Visions for sustainability. University of Turin, Torino, Italy

Goonesekere S (1998) A rights-based approach to realizing gender equality. United Nations Division for the Advancement of Women. www.un.org/womenwatch/daw/news/savitri.htm
Global Climate Risk Index (2021) Who suffers most from extreme weather events? Weather-related loss events in 2019. https://www.researchgate.net/publication/348759719_GLOBAL_CLIMATE_RISK_INDEX_2021
Halvard B, Gleditsch NP, Theisen OM (2008). Implications of climate change for armed conflict. World Bank, Washington, DC
IASC Reference Group (2018) Gender and humanitarian action. The Gender Handbook for Humanitarian Action. EU
Islam MR, Shamsuddoha M (2017) Socioeconomic consequences of climate induced human displacement and migration in Bangladesh. Int Sociol 32(3):277–298
Kuele G, Miola AC (2017) Somalia: the role of climate change in recurring violence. Igarapé Institute website
Lind A (2011) Women's community organizing in Quito: the paradoxes of survival and struggle. In: Visvanathan N et al (eds) The women, gender and development reader, no (2). Zed Books, London, pp 417–424
Mack A (2005) Human security report 2005. In: War and peace in the 21st century. Oxford Unit Press
Manchanda R (2021) "4: Difficult encounters with the WPS agenda in South Asia: re-scripting globalized norms and policy frameworks for a feminist peace". In: new directions in women, peace and security. Bristol University Press, Bristol, UK. Retrieved Jul 10, 2023, from https://doi.org/10.51952/9781529207774.ch004
Marin A, Naess LO (2017) Climate change adaptation through humanitarian aid? Promises, perils, and potentials of "new humanitarianism. In: Marin A et al (eds) Courting catastrophe? Humanitarian policy and practice in a changing climate, vol 48, no 4. IDS Bulletin, pp 15–30
Mawby B, Applebaum A (2021) Addressing future fragility: women, climate change and migration. In: Basu S et al (eds) New directions in women peace and security. Bristol University Press, UK, Bristol, pp 207–222
McCann BT, Davis JM, Osborne D, Durham C, O'Brien M, Raymond NA (2021) Quantifying climate change-relevant humanitarian programming and spending across five countries with high vulnerability to disaster. Disasters 45(4):819−843 (Overseas Development Institute)
Miall H (2003) Global Governance and Conflict Prevention. In: Cochrane, F., Duffy, R., Selby, J. (eds) Global Governance, Conflict and Resistance. Palgrave Macmillan, London. https://doi.org/10.1057/9781403943811_4
National Adaptation Plan of Bangladesh (2022) Ministry of environment forest and climate change. People's Republic of Bangladesh, Dhaka
National Strategy for Water and Sanitation Hard to Reach Areas of Bangladesh (2011) Local Government Division. Ministry of Local Government and Rural Development People's Republic of Bangladesh
Ní Aoláin FN, Haynes DF, Cahn N (2011) On the frontlines: gender, war, and the post-conflict process. Oxford University Press, Oxford
Nussbaum M (2000) Women and the human right to development: the capabilities approach. Cambridge University Press
OCHA (2016) Humanitarian development nexus. OCHA website. https://www.unocha.org/es/themes/humanitarian-development-nexus
OCHA (2021) Humanitarian response plan: Somalia. Humanitarian programme cycle 2021. https://reliefweb.int/sites/reliefweb.int/files/resources/HRP_2021-Somalia.pdf
OECD (2020) Learning from crises, building resilience. Development Co-operation Report 2020. OECD Publishing, Paris
Paffenholz T (2004) Designing transformation and intervention processes. Bergh of Research Centre for Constructive Conflict Management
Pankhurst D (2000) Women, gender and peacebuilding. In: Working Paper 5: Centre for Conflict Resolution. UK Department of Peace Studies, University of Bradford

Pankhurst D (2008) Post-war backlash violence against women: what can masculinity explain?. In: Pankhurst D (ed) Gendered peace: women's struggles for post-war justice and reconciliation. Routledge
Parlevliet M (2011) Human rights and conflict transformation. In: Towards a more integrated approach.
Paul D, Gleditsch NP (eds) (2018) Environmental conflict: an anthology. Routledge, New York
Peters K, Dupar M, Opitz-Stapleton S, Lovell E, Budimir M, Brown S, Cao Y (2020) Climate change, conflict and fragility: an evidence review and recommendations for research and action. Practical Action, London, UK
Puechguirbal N (2010) Discourses on gender, patriarchy and resolution 1325: a textual analysis of UN documents. Int Peacekeeping 17(2)
Rehn E, Johnson Sirleaf E (2002) Women war peace: the independent experts' assessment on the impact of armed conflict on women and women's role in peace-building. In: Progress of the world's women 2002, vol 1. UNIFEM, New York, USA
Sen A (1998) Development as freedom. Oxford University Press
Sharifi A, Simangan D, Kaneko S (2023) Integrated approaches to peace and sustainability. In: World sustianbility series. Springer
Simangan D, Sharifi A, Kaneko S (2021) Positive peace pillars and sustainability dimensions: an analytical framework. Analytical essay. Int Stud Rev (0):1–22 (Science Direct)
Simangan D, Lee CY, Sharifi A, Candelaria JL, Kaneko S (2022) A global analysis of interactions between peace and environmental sustainability. Earth Syst Gov (14) (Science Direct)
Simangan D, Bose S, Candelaria JL, Krampe F, Kaneko S (2023) Positive peace and environmental sustainability: local evidence from Afghanistan and Nepal. Environ Secur 1–21 (Sage)
Smith ES (2020) Climate change in women, peace and security national action plans. Stockholm International Peace Research Institute. Jstore
Stoakes E (2019) Leak suggests UN agency self-censors on climate crisis after US pressure. The Guardian. Accessed 11 July 2023
Swaine A, Spearing M, Murphy M, Contreras-Urbina M (2019) Exploring the intersection of violence against women and girls with post-conflict state building and peacebuilding processes: a new analytical framework. J Peacebuild Dev 14(1):3–21
Tompkins E, K. Vincent, R.J. Nicholls, and N. Suckall (2018) Documenting the state of adaptation for the global stocktake of the Paris Agreement. Wiley Interdiscip Rev: Clim Change 9(5)
United Nations Report of the World Conference Report (1980). United Nations Decade for Women. Chapter. 1. Section(A).
UNHCR (2021) Rohingya Refugee Response—Bangladesh Factsheet—Public Health. Cox's Bazar. UNHCR Bangladesh
United Nations Development Fund for Women (UNIFEM) (2007) CEDAW and the human rights based approach to programming. A UNIFEM Guide
United Nations Report of the World Conference (1980) United Nations Decade for Women. Chapter. 1. Section(A)
United Nations Report (2000) (UNSCR S/RES/1325). https://peacemaker.un.org/node/105, http://www.un.org/events/res_1325e.pdf. Accessed 2 Apr 2020
United Nations Report (2002) Women, Peace and Security Study. Secretary-General pursuant to Security Council resolution 1325 (2000)
United Nations Security Council Resolution Report (2019, 2020) UN Documents for Women Peace and Security. https://www.securitycouncilreport.org/un-documents/women-peace-and-security/. Accessed 10 July 2023
United Nations (1992) Report: an agenda for peace: preventive diplomacy, peacemaking and peace-keeping
United Nations (2023) Sustainable development goals. https://www.un.org/sustainabledevelopment/peace-justice/
UNDP Gender and Climate Finance Report (2017) Issue brief. UNDP. https://www.undp.org/publications/gender-and-climate-finance

United Nations General Assembly Resolution 62/136 (2008) Proclaiming the International Day of Rural Women
UN General Assembly Resolution (2008) Report. UNGA, New York. Accessed: https://www.un.org/dppa/decolonization/en/fourth-committee/63rd-session-2008#:~:text=During%20the%2063rd%20session,General%20Assembly%20(A%2FC
Van Den Hoek J, Friedrich HK, Ballasiotes A, Peters LER, Wrathall D (2021) Development after displacement: evaluating the utility of open street map data for monitoring sustainable development goal progress in refugee settlements. ISPRS Int
Yoshida K (2019) The nature of women peace and security: where is the environment in WPS and where is WPS in environmental peacebuilding? Centre for Women Peace and Security Research. London School of Economics, London
Yousuf F (2022) Making visible the afghan men who are working for women's rights and a gender-just society. SwissPeace
World Food Programme (WFP) (2022) Impact assessment of the WFP's resilience building activities in the Rohingya refugee camps, Cox's Bazar. Evaluation Report. WFP, Cox's Bazar
Women, Peace, and Security Index (2021) Summary: tracking sustainable peace through inclusion, justice, and security for women. Georgetown Institute for Women, Peace and Security and Peace Research Institute Oslo, Washington, DC

MIX
Papier aus verantwortungsvollen Quellen
Paper from responsible sources
FSC® C105338

If you have any concerns about our products,
you can contact us on
ProductSafety@springernature.com

In case Publisher is established outside the EU,
the EU authorized representative is:
Springer Nature Customer Service Center GmbH
Europaplatz 3, 69115 Heidelberg, Germany

Printed by Libri Plureos GmbH
in Hamburg, Germany